Letts

KS3
Success

Science

Workbook

Age 11-14

Dan Foulder

CONTENTS

Energy

Motion and forces

Waves

Electricity and electromagnetism

Matter and space physics

A Choose just one answer: a, b, c or d.

1 What device is required for viewing cells? (1 mark)
  a) a telescope ◯
  b) an MRI scanner ◯
  c) a magnifying glass ◯
  d) a light microscope ◯

2 What is the function of the nucleus? (1 mark)
  a) photosynthesis ◯
  b) aerobic respiration ◯
  c) anaerobic respiration ◯
  d) controls the actions of the cell ◯

3 Which of the following statements best describes the cell membrane? (1 mark)
  a) controls the cell ◯
  b) surrounds the cell and allows material to enter and exit ◯
  c) where photosynthesis occurs ◯
  d) only found in plant cells ◯

4 Which of these structures are found in both animal and plant cells? (1 mark)
  a) mitochondria ◯   **b)** chloroplasts ◯
  c) vacuoles ◯   **d)** cell walls ◯

5 What term describes an organism that consists of only one cell? (1 mark)
  a) polycellular ◯   **b)** multicellular ◯
  c) monocellular ◯   **d)** unicellular ◯

Score   **/5**

B Answer each question.

1 What are the fundamental units of living organisms? (1 mark)

....

2 By what process do materials cross the cell membrane? (1 mark)

....

3 What is the name of the jelly-like fluid where chemical reactions occur inside the cell? (1 mark)

....

4 What name is given to a group of similar cells that are working together to carry out the same function? (1 mark)

....

5 What do groups of organs that work together form? (1 mark)

....

6 What term is used to describe an organism that is made up of more than one cell? (1 mark)

....

Score   **/6**

C   **Answer all parts of the questions. Use a separate sheet of paper if necessary.**

1 The diagram shows a key used to identify different parts of cells.

Give the names of parts A–E.   (5 marks)

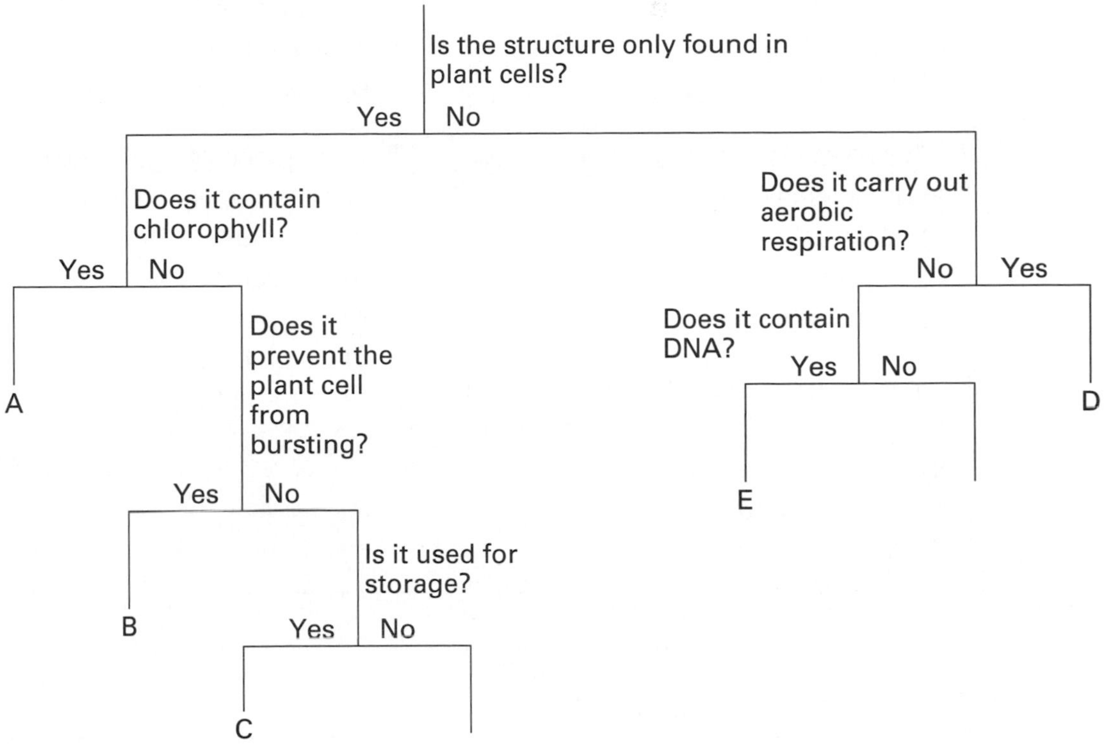

Is the structure only found in plant cells?

       Yes | No

Does it contain chlorophyll?

   Yes | No

A

Does it prevent the plant cell from bursting?

   Yes | No

B

Is it used for storage?

   Yes | No

C

Does it carry out aerobic respiration?

     No | Yes

Does it contain DNA?

   Yes | No

E

D

A ....

B ....

C ....

D ....

E ....

2 Put the following levels of organisation in order of complexity, with the least complex first.

     Organ system   **Organ**   **Cell**   **Tissue**   (2 marks)

....

....

Score   **/7**

For more help on this topic see KS3 Science Revision Guide pages 4–5.

A Choose just one answer: a, b, c or d.

1 What is the skeleton composed of? (1 mark)
   a) bones ○
   b) muscle ○
   c) ligaments ○
   d) skin ○

2 What attaches muscles to bone? (1 mark)
   a) ligaments ○
   b) skin ○
   c) tendons ○
   d) small bones ○

3 What are muscles able to do? (1 mark)
   a) push ○
   b) pull ○
   c) push and pull ○
   d) push then pull ○

4 What joins bones to other bones? (1 mark)
   a) tendons ○
   b) muscles ○
   c) ligaments ○
   d) cellulose ○

5 What piece of equipment can be used to measure the force exerted by muscles? (1 mark)
   a) a nanometer ○
   b) a gyroscope ○
   c) a forcemeter ○
   d) a force trigger ○

Score /5

B Answer each question.

1 What happens to a muscle when it contracts? (2 marks)

....

....

2 What happens to a muscle when it relaxes? (1 mark)

....

3 What is an antagonistic pair of muscles? (1 mark)

....

....

4 When one of an antagonistic pair of muscles is contracting, what is the other muscle doing? (1 mark)

....

Score /5

1 The functions of the skeleton are shown below.

Protection   Movement   Production of blood cells   Support

Write the correct function next to the description in the table below.   (4 marks)

Function	Description
	Provides an anchor point for muscles
	Prevents the organs from being damaged
	Cells are produced in the bone marrow
	Provides structure

2 Lee investigates the force generated by different muscles. Lee's results are shown below.

Muscle	Force (N)
Bicep	250
Tricep	200
Deltoids	170
Calf muscles	300

a) On the graph grid below plot a bar chart of this data.   (4 marks)

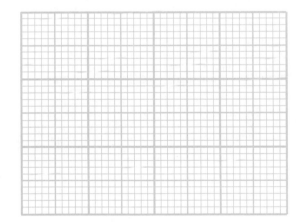

b) Lee carried out the investigation twice more to get another two sets of results. This allowed him to calculate mean results for each muscle. What is the benefit of doing this?   (1 mark)

....

....

Score   /9

For more help on this topic see KS3 Science Revision Guide pages 6–7.

A Choose just one answer: a, b, c or d.

1 Which health problem can be caused by not having enough to eat? (1 mark)
   a) obesity ○
   b) starvation ○
   c) Crohn's disease ○
   d) emphysema ○

2 What substance helps move undigested food through the digestive system? (1 mark)
   a) protein ○   b) vitamins ○
   c) roughage ○   d) fats ○

3 Fats and oils are types of: (1 mark)
   a) carbohydrates ○   b) proteins ○
   c) lipids ○   d) vitamins ○

4 What is the function of fat in the body? (1 mark)
   a) energy storage ○
   b) transporting substances in the blood ○
   c) repair ○
   d) to prevent rickets ○

5 What is the function of proteins in the body? (1 mark)
   a) energy storage ○
   b) move undigested food through the gut ○
   c) preventing obesity ○
   d) growth and repair ○

Score   /5

B Answer each question.

1 Explain the importance of carbohydrates in the diet. (2 marks)

....

....

2 Why are enzymes important in digesting food? (2 marks)

....

....

3 What are the benefits of gut bacteria (bacteria which live in the digestive system of humans)? (2 marks)

....

....

4 What condition can lead to health problems such as heart disease and Type 2 diabetes? (1 mark)

....

5 Why is the liver important in digestion? (2 marks)

....

Score   /9

NUTRITION AND DIGESTION

MODULE 3

1 The pie chart below shows the components of a balanced diet.

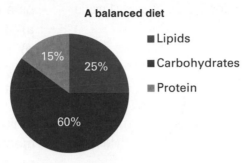

A balanced diet

- Lipids
- Carbohydrates
- Protein

15% 25% 60%

The pie charts A–C show three different diets.

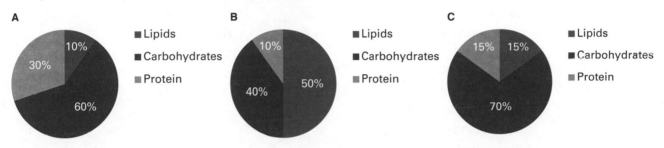

A
- Lipids
- Carbohydrates
- Protein

10% 30% 60%

B
- Lipids
- Carbohydrates
- Protein

10% 40% 50%

C
- Lipids
- Carbohydrates
- Protein

15% 15% 70%

a) Which diet would be most likely to lead to obesity? Explain your answer. (3 marks)

....

....

....

b) Vitamins and minerals are not shown on the charts but they are required by the body. Name two conditions that can be caused by not having enough vitamins and minerals in the diet. (1 mark)

....

2 The diagram shows the human digestive system.

Complete the table with the names of the organs A–C and their functions in digestion. (3 marks)

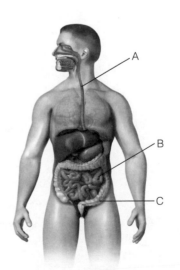

A

B

C

	Organ name	Function
A		
B		
C		

Score /7

For more help on this topic see KS3 Science Revision Guide pages 8–9.

A Choose just one answer: a, b, c or d.

1 Which structure connects the trachea to the bronchioles? (1 mark)
a) epiglottis ○
b) duodenum ○
c) lumen ○
d) bronchi ○

2 Which device can be used to measure a person's vital capacity? (1 mark)
a) spirometer ○
b) calorimeter ○
c) colorimeter ○
d) nanometer ○

3 Which structure allows gases to enter a leaf? (1 mark)
a) endodermis ○
b) palisade mesophyll ○
c) stomata ○
d) cortex ○

4 What effect will regular exercise have on a person's vital capacity? (1 mark)
a) it will stay the same ○
b) it will increase ○
c) it will decrease slightly ○
d) it will decrease significantly ○

5 Which of the following terms means 'breathing in'? (1 mark)
a) excretion ○
b) inspiration ○
c) expiration ○
d) egestion ○

Score /5

B Answer each question.

1 What is the function of the gas exchange system in humans? (3 marks)

....

....

....

2 Why is it important that the volume in the lungs is decreased when breathing out? (2 marks)

....

....

3 What causes a person to find it difficult to breathe during an asthma attack? (2 marks)

....

....

4 How is asthma normally treated? (2 marks)

....

....

Score /9

1 The diagram below shows the gas exchange system in humans.

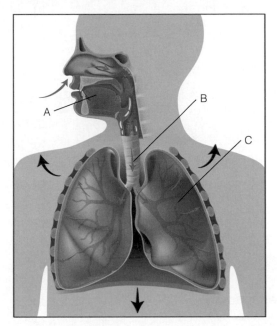

Complete the table with the names of the organs A–C and their functions in gas exchange. (3 marks)

	Organ name	Function
A		
B		
C		

2 a) The number of cases of asthma in the UK has risen over the last 50 years. The number of cars on the road has also increased in the same time period. Josh thinks this proves that increased pollution from cars is leading to an increase in the cases of asthma.

Why might Josh think this? (2 marks)

....

....

b) Amy thinks this data does not prove that increased pollution from cars is causing an increase in the number of cases of asthma. Do you agree with her?

Give reasons for your answer. (3 marks)

....

....

....

Score /8

For more help on this topic see KS3 Science Revision Guide pages 10–11.

A Choose just one answer: a, b, c or d.

1 What is fertilisation? (1 mark)
- a) the penis entering the vagina ○
- b) the egg being released from the ovary ○
- c) when the sperm fuses with the egg ○
- d) when the embryo implants ○

2 What is the male gamete in humans? (1 mark)
- a) sperm ○
- b) egg ○
- c) pollen ○
- d) zygote ○

3 When does the menstrual cycle begin in women? (1 mark)
- a) birth ○
- b) the menopause ○
- c) puberty ○
- d) six months after birth ○

4 What is the endometrium? (1 mark)
- a) where sperm is stored ○
- b) the inner layer of the ovary ○
- c) the bottom of the cervix ○
- d) the lining of the uterus ○

5 What is the function of the male gamete's tail? (1 mark)
- a) it provides protection from predators ○
- b) it allows the male gamete to crawl to the egg ○
- c) it allows the male gamete to swim to the egg ○
- d) signalling ○

Score /5

B Answer each question.

1 If a mother drinks or takes drugs whilst pregnant it can cause damage to the developing foetus. Explain why. (2 marks)

....

....

....

2 Where does fertilisation occur in humans? (2 marks)

....

....

3 Where are sperm cells produced in humans? (2 marks)

....

....

4 Where does the development of the embryo occur? (2 marks)

....

Score /8

C | **Answer all parts of the questions. Use a separate sheet of paper if necessary.**

1 Number the following stages 1–6 to give the correct sequence of events for reproduction to occur in humans. The first one has been done for you. (4 marks)

● The zygote divides to form an embryo then moves down the oviduct to the uterus where it implants in the endometrium. ☐

● After nine months the mother will go into labour. Muscle contractions push the foetus out of the vagina and the baby is born. ☐

● Once a month, an egg is released from the ovaries and the lining of the uterus (the endometrium) thickens in preparation for implantation. 1

● If fertilisation does take place the sperm will meet the egg in the oviduct. ☐

● The embryo develops into a foetus and is provided with nutrients from the mother's blood by the placenta. ☐

● The sperm and egg fuse to form a zygote. ☐

2 Explain why, in humans, the egg cell is much larger than the sperm cell. (2 marks)

....

....

3 Explain why menstruation does not occur when a woman is pregnant. (2 marks)

....

....

....

....

Score **/8**

For more help on this topic see KS3 Science Revision Guide pages 12–13.

A — Choose just one answer: a, b, c or d.

1 Which of the following statements is incorrect? (1 mark)
a) plants can reproduce sexually ◯
b) plant reproduction can involve pollen ◯
c) plants cannot reproduce sexually ◯
d) plants are not able to reproduce ◯

2 What are the reproductive organs of a plant? (1 mark)
a) roots ◯   b) stems ◯
c) leaves ◯   d) flowers ◯

3 What does the fertilised zygote in a plant form? (1 mark)
a) flowers ◯   b) seeds ◯
c) fruits ◯   d) vines ◯

4 What do fruits contain? (1 mark)
a) seeds ◯
b) pollen ◯
c) flowers ◯
d) muscle ◯

5 What is pollination? (1 mark)
a) transfer of pollen from the anther to the stigma ◯
b) transfer of sperm from the stigma to the anther ◯
c) transfer of pollen from the stigma to the anther ◯
d) transfer of sperm from the anther to the stigma ◯

Score /5

B — Answer each question.

1 What is the function of a pollen tube in a plant? (2 marks)

2 Explain the difference between the flowers of an insect-pollinated plant and a wind-pollinated plant. (3 marks)

3 Why would having small hooks on a seed help it to be dispersed? (2 marks)

Score /7

Answer all parts of the question. Use a separate sheet of paper if necessary.

1 The map below shows the positions of two different plant species, A and B. Seeds from plant A have been found in area 1 but not in area 2. Seeds from plant B have been found in areas 1 and 2.

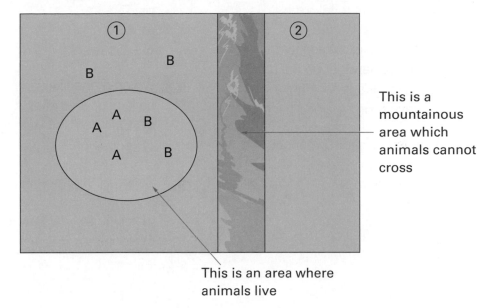

This is a mountainous area which animals cannot cross

This is an area where animals live

a) i) What is the most likely method of seed dispersal for plant A? (1 mark)

....

ii) Explain your answer. (1 mark)

....

....

b) i) What is the most likely method of seed dispersal for plant B? (1 mark)

....

ii) Explain your answer. (1 mark)

....

....

c) Insects are unable to cross the mountainous region. However, members of plant species B from areas 1 and 2 are still able to interbreed. Explain the most likely reason for this. (2 marks)

....

....

....

Score **/6**

For more help on this topic see KS3 Science Revision Guide pages 14–15.

A Choose just one answer: a, b, c or d.

1 When a person becomes addicted to a drug they: (1 mark)
a) don't want to take the drug anymore ◯
b) have never taken the drug ◯
c) are dependent on the drug ◯
d) have taken the drug once but have then stopped taking it ◯

2 What is cannabis produced from? (1 mark)
a) an insect ◯
b) a fungus ◯
c) crude oil ◯
d) a plant ◯

3 Which of the following is an extremely addictive opiate which suppresses pain? (1 mark)
a) LSD ◯   b) ecstasy ◯
c) cocaine ◯   d) heroin ◯

4 Which of the following drugs has hallucinogenic effects? (1 mark)
a) LSD ◯   b) ecstasy ◯
c) cocaine ◯   d) tobacco ◯

5 Which of the following drugs is illegal? (1 mark)
a) cocaine ◯   b) alcohol ◯
c) tobacco ◯   d) caffeine ◯

Score /5

B Answer each question.

1 Why are heroin users who inject the drug at a higher risk of contracting HIV? (1 mark)

....

2 What does the term 'solvent abuse' refer to? (1 mark)

....

3 What is the addictive chemical in tobacco? (1 mark)

....

4 What chemical in tobacco smoke increases the risk of users developing lung cancer? (1 mark)

....

5 Why might the use of LSD lead to accidents? (2 marks)

....

....

....

Score /6

Answer all parts of the questions. Use a separate sheet of paper if necessary.

1 The graph below shows the number of deaths from liver disease in the UK over time.

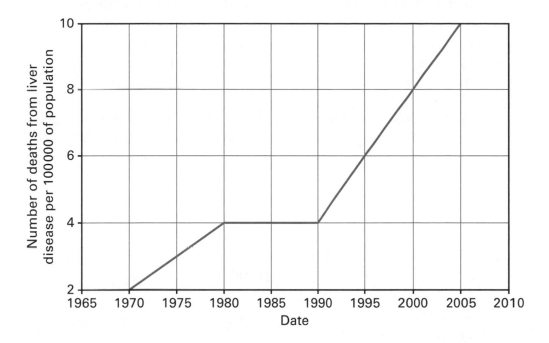

a) Describe the trend shown by the graph. (3 marks)

....

....

....

b) Would banning alcoholic drinks reduce the number of deaths from liver disease?

Explain your answer. (2 marks)

....

....

....

2 Why is drinking alcohol before driving considered dangerous? (3 marks)

....

....

....

....

Score   /8

For more help on this topic see KS3 Science Revision Guide pages 16–17.

A   Choose just one answer: a, b, c or d.

1 Which of the organisms below
can photosynthesise?   (1 mark)
- a) birds ⃝
- b) plants ⃝
- c) fungi ⃝
- d) insects ⃝

2 Where does photosynthesis occur
in a plant cell?   (1 mark)
- a) nucleus ⃝
- b) cytoplasm ⃝
- c) mitochondria ⃝
- d) chloroplasts ⃝

3 What chemical is used in
photosynthesis?   (1 mark)
- a) chlorophyll ⃝
- b) ADH ⃝
- c) NAD ⃝
- d) florigen ⃝

4 What term describes the permanent
removal of trees?   (1 mark)
- a) draining ⃝
- b) aforestation ⃝
- c) deforestation ⃝
- d) rotation ⃝

5 What are the main photosynthetic
organs of plants?   (1 mark)
- a) roots ⃝
- b) stems ⃝
- c) flowers ⃝
- d) leaves ⃝

Score   **/5**

B   Answer each question.

1 What are the reactants in photosynthesis?   (2 marks)

....

2 What are the products of photosynthesis?   (2 marks)

....

3 What provides the energy for photosynthesis?   (1 mark)

....

4 How could the large-scale removal of trees affect the Earth's climate?   (3 marks)

....

....

....

Score   **/8**

MODULE 8   PHOTOSYNTHESIS

C Answer all parts of the question. Use a separate sheet of paper if necessary.

1 Sidra and James are investigating photosynthesis in different unicellular plant cells..

They prepare four different samples (A–D), which contain unicellular plant cells. They then change the conditions in each of the tubes as shown in the table below.

Sample	Carbon dioxide present	Water present	Light	Chloroplasts present in plant cells
A	✓	✓	✗	✓
B	✗	✓	✓	✓
C	✗	✓	✓	✗
D	✓	✓	✓	✓

a) For each sample, predict if the plant cells will photosynthesise and explain why photosynthesis does or does not occur. (4 marks)

Sample	Photosynthesis occurs	Explanation
A		
B		
C		
D		

b) The volume of a gas produced over time could be used as a way of measuring the rate of photosynthesis in this investigation.

What gas would be used? Explain your answer. (2 marks)

....

....

....

Score /6

For more help on this topic see KS3 Science Revision Guide pages 22–23.

A — Choose just one answer: a, b, c or d.

1 Which process releases energy from food? (1 mark)
- a) photosynthesis ○
- b) accumulation ○
- c) respiration ○
- d) absorption ○

2 Which gas is required for anaerobic respiration to occur? (1 mark)
- a) nitrogen ○
- b) oxygen ○
- c) carbon dioxide ○
- d) no gas is required for anaerobic respiration ○

3 During which process is lactic acid produced? (1 mark)
- a) photosynthesis ○
- b) anaerobic respiration in humans ○
- c) anaerobic respiration in microorganisms ○
- d) digestion ○

4 What is another name for anaerobic respiration in microorganisms? (1 mark)
- a) alcoholic fermentation ○
- b) photosynthesis ○
- c) deamination ○
- d) digestion ○

5 Where does aerobic respiration occur? (1 mark)
- a) mitochondria ○
- b) chloroplasts ○
- c) Golgi body ○
- d) nucleus ○

Score /5

B — Answer each question.

1 What are the reactants in aerobic respiration? (1 mark)

....

2 What are the products of aerobic respiration? (1 mark)

....

3 What is the reactant in anaerobic respiration? (1 mark)

....

4 What are the products of anaerobic respiration in microorganisms? (1 mark)

....

5 What is the product of anaerobic respiration in animals? (1 mark)

....

Score /5

CELLULAR RESPIRATION

MODULE 9

1 Harry and Aisha carry out an investigation into anaerobic respiration.

Their apparatus is shown below.

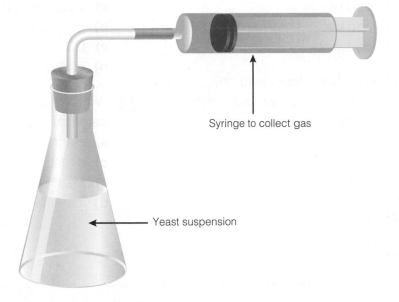

Syringe to collect gas

Yeast suspension

a) What must they do to ensure anaerobic conditions? (1 mark)

....

b) What gas is being collected in the syringe? (1 mark)

....

c) Could the same apparatus be used to investigate anaerobic respiration in a sample
of human cells? Explain your answer. (3 marks)

....

....

....

d) Aisha wants to carry out a different investigation into aerobic respiration using woodlice.
Harry is concerned about the ethics of this new investigation.

What concerns could Harry have? (1 mark)

....

....

Score **/6**

For more help on this topic see KS3 Science Revision Guide pages 24–25.

A Choose just one answer: a, b, c or d.

1 The food chains in an ecosystem can be combined to form a: (1 mark)
a) food matrix ○
b) food nexus ○
c) food net ○
d) food web ○

2 What do herbivores feed on? (1 mark)
a) insects ○ b) mammals ○
c) plants ○ d) birds ○

3 How is energy lost in food chains? (1 mark)
a) heat ○
b) UV radiation ○
c) light ○
d) growth ○

4 Which term describes a pesticide that has long-lasting effects in an ecosystem? (1 mark)
a) instantaneous ○
b) persistent ○
c) DDT ○
d) conservation ○

5 Which of the following is not eaten by a carnivore? (1 mark)
a) plants ○
b) birds ○
c) mammals ○
d) insects ○

Score /5

B Answer each question.

1 Why are pesticides used on farms? (2 marks)

....

....

....

....

2 Describe how toxic materials that are taken in by primary consumers can affect other organisms. (2 marks)

....

....

....

3 Describe how the use of pesticides could affect the food security of humans. (3 marks)

....

....

....

....

Score /7

Answer all parts of the question. Use a separate sheet of paper if necessary.

1 The paragraph below describes the feeding relationships in a marine ecosystem.

Phytoplankton are very small algae that photosynthesise. They are fed on by small organisms called zooplankton. Mussels also feed on phytoplankton and zooplankton. Fish eat zooplankton and fish are eaten by seabirds. The seabirds are also able to eat mussels.

a) Use the information above to draw a food web to show the feeding relationships in this ecosystem. (5 marks)

b) Use your food web to give an example of:

i) a primary producer (1 mark)

....

ii) a primary consumer (1 mark)

....

iii) a secondary consumer (1 mark)

....

Score /8

For more help on this topic see KS3 Science Revision Guide pages 26–27.

A Choose just one answer: a, b, c or d.

1 Which molecule carries genetic information? (1 mark)
a) ATP ○ b) glucose ○
c) ADH ○ d) DNA ✓

2 Which term is given to the differences between organisms of the same species? (1 mark)
a) sterility ○
b) variation ✓
c) evolution ○
d) competition ○

3 Which term describes the point at which all members of a species die out? (1 mark)
a) deforestation ○
b) accumulation ○
c) conservation ○
d) extinction ✓

4 Which of the following scientists was **not** involved in the discovery of the structure of DNA? (1 mark)
a) Hooke ○
b) Watson ○
c) Franklin ○
d) Crick ✓

5 Which term is used for the number of different species living in an area? (1 mark)
a) ecology ○
b) conservation ○
c) biodiversity ✓
d) bioaccumulation ○

Score   /5

B Answer each question.

1 Explain how gene banks can help conservation. (3 marks)

....

....

....

....

2 What is 'heredity'? (2 marks)

....

....

3 Is eye colour an example of continuous variation or discontinuous variation? Explain your answer. (2 marks)

....

....

Score   /7

1 Label the diagram below with the following key words. (2 marks)

cell　　　　　**DNA**　　　　　**chromosome**

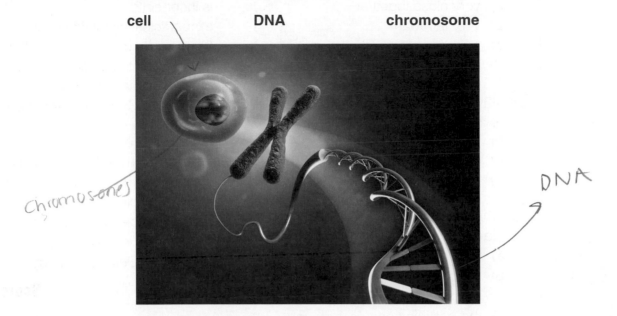

chromosomes　　　　　　　　　　　　　　　　*DNA*

2 Two insect species have very similar appearances.

Mariyah and David are studying four different populations (A–D) of insects but they are not sure which populations are the same species and which are not.

They breed the different populations together and their results are shown in the table below.

'Yes' indicates that the insects produced fertile offspring and 'No' indicates that they did not.

	A	B	C	D
A	Yes	No	No	Yes
B	No	No	No	No
C	No	No	No	No
D	Yes	No	No	Yes

a) Which populations were the same species? (1 mark)

....

b) Explain how you arrived at your answer for part **a)**. (2 marks)

....

....

....

Score **/5**

For more help on this topic see KS3 Science Revision Guide pages 28–29.

A Choose just one answer: a, b, c or d.

1 The particles in solids are: (1 mark)
a) very close together ○
b) very far apart ○
c) quite far apart ○
d) never able to touch each other ○

2 Liquids are able to: (1 mark)
a) expand to fill any space ○
b) flow ○
c) maintain their shape ○
d) be compressed easily ○

3 Which of the following statements about the particles in a gas is correct? (1 mark)
a) the particles are free to move ○
b) the particles are only able to vibrate ○
c) the particles can never touch the sides of the container they are in ○
d) the particles have very strong attractive forces between them ○

4 Which of the following statements is incorrect? (1 mark)
a) particles in a solid vibrate ○
b) there are strong attractive forces between the particles in a solid ○
c) solids have a fixed volume ○
d) particles in a solid are unable to move ○

5 What happens to the particles of a substance when it is heated? (1 mark)
a) the particles move slower ○
b) the particles move faster ○
c) the particles become stationary ○
d) the particles lose energy ○

Score /5

B Answer each question.

1 Describe how the pressure of a gas can be increased. (2 marks)

2 Describe how the pressure of a gas can be decreased. (2 marks)

3 Describe what happens to a particle when it is heated. (2 marks)

Score /6

C **Answer all parts of the question. Use a separate sheet of paper if necessary.**

1 **a)** For each of the following processes, explain why the state of matter given is the most suitable for the purpose.

   i) Solids are used to construct the walls of a house. (2 marks)

   ....

   ....

   ii) Gas is used to fill hot air balloons. (2 marks)

   ....

   ....

   iii) Liquids are used to fill swimming pools. (2 marks)

   ....

   ....

b) Use your knowledge of particles to explain the change in pressure when the gas in a hot air balloon is heated. (3 marks)

....

....

....

c) Cars have hydraulic brake systems. When the driver presses on the brake pedal, the matter in the brake tubes transfers the force to the brakes. In order to do this, the matter in the tube must be able to flow and be difficult to compress.

What state of matter is used in brake tubes? Explain your answer. (3 marks)

....

....

....

Score **/12**

For more help on this topic see KS3 Science Revision Guide pages 32–33.

A Choose just one answer: a, b, c or d.

1 What will happen to a liquid that is cooled? (1 mark)
a) it will expand ○
b) it will contract ○
c) it will evaporate ○
d) it will melt ○

2 What will happen to a solid that is heated? (1 mark)
a) it will expand ○
b) it will contract ○
c) it will not change ○
d) it will freeze ○

3 What will happen to a liquid when it is heated above its boiling point? (1 mark)
a) the particles will move slower ○
b) its shape will become constant ○
c) it will form a solid ○
d) it will form a gas ○

4 How can a solid be turned into a liquid? (1 mark)
a) heat it above its melting point ○
b) heat it above its boiling point ○
c) freeze it ○
d) increase the energy of the particles ○

5 What will happen to a gas that is cooled below its boiling point? (1 mark)
a) the particles will move slower ○
b) its volume will not be constant ○
c) it will remain as a gas ○
d) the particles will move faster ○

Score /5

B Answer each question.

1 What happens to a liquid that is cooled below its freezing point?

Describe what happens to the particles in the liquid in your answer. (3 marks)

....

....

....

....

2 What happens to a solid that is heated past its melting point?

Describe what happens to the particles in the solid in your answer. (3 marks)

....

....

....

....

Score /6

EXPANSION AND CONTRACTION

MODULE 13

1 An experiment investigated the contraction of different metals when they were cooled. The results are shown below.

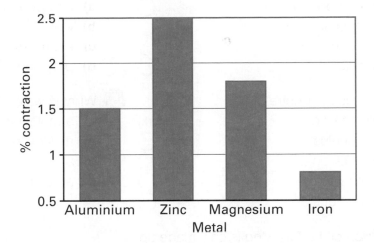

a) Which of the metals shown in the graph:

   i) contracted the most? (1 mark)

   ....

   II) contracted the least? (1 mark)

   ....

b) What would cause the metals to expand? (1 mark)

   ....

c) Liquids also expand and contract.

   Explain how this is used in a thermometer to determine temperature. (4 marks)

   ....

   ....

   ....

   ....

   ....

   Score   /7

For more help on this topic see KS3 Science Revision Guide pages 34–35.

A — Choose just one answer: a, b, c or d.

1 What is the simplest unit of matter? (1 mark)
- a) compound ○
- b) element ○
- c) atom ○
- d) photon ○

2 What is the centre of an atom called? (1 mark)
- a) nucelus ○
- b) nucleon ○
- c) nucleolus ○
- d) nucleus ○

3 Which of the following is only made up of one type of atom? (1 mark)
- a) a compound ○
- b) an acid ○
- c) an alkali ○
- d) an element ○

4 Which of the following are organised in the Periodic Table? (1 mark)
- a) compounds ○
- b) ions ○
- c) elements ○
- d) polymers ○

5 What is water (H_2O)? (1 mark)
- a) a molecule ○
- b) a single atom ○
- c) an ion ○
- d) an element ○

Score /5

B — Answer each question.

1 What is a compound? (1 mark)

....

....

2 What happens to the mass of a substance when it melts? (1 mark)

....

3 What happens to the mass of a substance when it boils? (1 mark)

....

4 In a chemical reaction, calcium carbonate decomposes to form calcium oxide and carbon dioxide. The total mass of the calcium oxide and carbon dioxide produced is 5g.

What was the mass of calcium carbonate at the start of the reaction? (1 mark)

....

Score /4

C **Answer all parts of the questions. Use a separate sheet of paper if necessary.**

1 a) Complete the table below with the following terms.

Some of the terms can be used more than once. (3 marks)

positive negative neutral $\dfrac{1}{1800}$ 1

Particle	Relative mass	Charge
Proton		
Neutron		
Electron		

b) Identify the parts of the atom labelled A, B and C. (3 marks)

A ....

B ....

C ....

2 Ali and Claire are investigating conservation of mass. They burn a sample of wood. As the wood is burned, smoke is produced. At the end of the investigation, the ash left over had a smaller mass than the original sample of wood. Ali believed this proved that mass is not always conserved in a reaction.

Is Ali correct? Explain your answer. (3 marks)

....

....

....

....

Score /9

For more help on this topic see KS3 Science Revision Guide pages 36–37.

A Choose just one answer: a, b, c or d.

1 Which of the following terms is used for two or more substances that are combined together but not chemically bonded? (1 mark)
a) compound ○ b) molecule ○
c) element ○ d) mixture ○

2 Diffusion involves movement from: (1 mark)
a) a low to a high concentration ○
b) a high to a low concentration ○
c) a high mass to a low mass ○
d) a low pressure to a high pressure ○

3 A pure substance contains: (1 mark)
a) many different types of atom ○
b) one type of atom or molecule ○
c) many different molecules ○
d) a mixture of different atoms and molecules ○

4 What is the function of the condenser in distillation? (1 mark)
a) cools the gas ○
b) evaporates the liquid ○
c) filters the solid ○
d) dissolves the solute ○

5 In what type of substance will diffusion occur fastest? (1 mark)
a) liquids ○
b) solids ○
c) gases ○
d) none of the above – the speed of diffusion is constant ○

Score /5

B Answer each question.

1 Describe how impurities may affect the melting and boiling point of a substance. (2 marks)

2 Name two separation methods. (2 marks)

3 What is a solute? (1 mark)

4 What is a solvent? (1 mark)

5 What is a solution? (1 mark)

Score /7

Answer all parts of the questions. Use a separate sheet of paper if necessary.

1 The three beakers below show the results of investigating the solubility of different substances in water.

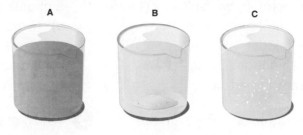

The table below shows the solubility of the different substances used.

Substance	Solubility
Copper sulfate (blue)	Very soluble
Salt	Quite soluble
Sand	Insoluble

a) Identify the substances in beakers A, B and C using the information in the table.   (3 marks)

A ....

B ....

C ....

b) If solution A is poured into a large tank of water, the blue colour spreads through the tank. What process causes the blue colour to spread?   (1 mark)

....

2 The diagram alongside shows the results of separating a dye.

a) What method has been used to separate chemicals A–C in the dye?   (1 mark)

....

b) Place the three different chemicals in the dye in order of solubility, starting with the most soluble.   (2 marks)

....

....

c) Explain your answer to part **b)**.   (1 mark)

....

....

Score   /8

For more help on this topic see KS3 Science Revision Guide pages 38–39.

CHEMICAL REACTIONS

MODULE 16

A Choose just one answer: a, b, c or d.

1 In a word equation, where are the products normally written? (1 mark)
a) on the left-hand side ◯
b) in the middle ◯
c) on the right-hand side ◯
d) beneath the arrow ◯

2 In a word equation, where are the reactants normally written? (1 mark)
a) on the left-hand side ◯
b) in the middle ◯
c) on the right-hand side ◯
d) beneath the arrow ◯

3 Which of the following is not a type of chemical reaction? (1 mark)
a) displacement ◯ b) freezing ◯
c) combustion ◯ d) oxidation ◯

4 Which of the following speeds up a chemical reaction? (1 mark)
a) hydrogen ◯
b) an alkali ◯
c) a catalyst ◯
d) argon ◯

5 Which of the following is an acidic pH? (1 mark)
a) 6 ◯
b) 7 ◯
c) 12 ◯
d) 9 ◯

Score /5

B Answer each question.

1 What type of reaction is a metal corroding in air? (1 mark)

2 What is shown by the pH scale? (1 mark)

3 What is the pH of water? (1 mark)

4 What is phenolphthalein? (1 mark)

5 Why are strong acids and alkalis a safety risk? (1 mark)

6 What equipment should always be worn when working with acids or alkalis? (1 mark)

Score /6

C   **Answer all parts of the questions. Use a separate sheet of paper if necessary.**

1 Jess investigates the reactions of an acid or alkali.

    a) What would you expect Jess' investigation to show? Complete the word equations below with the following options.   (2 marks)

         weak acid   **strong alkali**   **neutral**   **weak alkali**

         weak acid + weak alkali → ....

         strong alkali + ... → weak alkali

         strong acid + ... → neutral

         .... + strong acid → weak acid

    b) What can Jess use to test the pH of the solutions she makes?   (1 mark)

         ....

2 Draw lines to join each description to the correct reaction.   (3 marks)

A substance is burned in air	thermal decomposition
A substance breaks down when heated	combustion
A reaction with oxygen to form an oxide	oxidation
An element replaces a more reactive element in a compound	displacement

3 Steve was carrying out an experiment. He added platinum to the reactants and the platinum was still present at the end of the reaction. When he repeated the experiment without using platinum, the reaction was much slower.

     What was platinum acting as in this reaction? Explain your answer.   (2 marks)

     ....

     ....

Score   **/8**

For more help on this topic see KS3 Science Revision Guide pages 40–41.

A Choose just one answer: a, b, c or d.

1 What will happen to the temperature of a substance that is gaining energy? (1 mark)
a) it will increase ◯
b) it will decrease ◯
c) it will increase then decrease ◯
d) it will decrease then increase ◯

2 What will happen to the temperature of a substance releasing energy into the environment? (1 mark)
a) it will increase ◯
b) it will decrease ◯
c) it will increase then decrease ◯
d) it will decrease then increase ◯

3 What will happen to the temperature of a solid as it melts? (1 mark)
a) it will increase ◯
b) it will decrease ◯
c) it will stay the same ◯
d) it will decrease then increase ◯

4 What type of reaction releases energy to the environment? (1 mark)
a) endothermic ◯
b) exothermic ◯
c) ectothermic ◯
d) ensothermic ◯

5 What type of reaction takes energy in from the environment? (1 mark)
a) endothermic ◯
b) exothermic ◯
c) ectothermic ◯
d) ensothermic ◯

Score /5

B Answer each question.

1 What happens to the temperature of an object as it freezes? (1 mark)

....

2 In what type of reaction would the temperature decrease as the reaction occurred? (1 mark)

....

3 In what type of reaction would the temperature increase as the reaction occurred? (1 mark)

....

4 Describe how the temperature of a substance will change as it is cooled as a gas and then formed as a liquid. (3 marks)

....

....

....

Score /6

Answer all parts of the question. Use a separate sheet of paper if necessary.

1 Kat and Temi heat a sample of liquid. They heat the liquid in a sealed container and record the temperature every five seconds. Their results are shown in the graph below.

The boiling point of the liquid was 110°C.

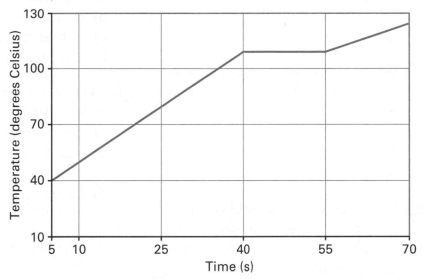

a) At what time did the sample begin to boil? (1 mark)

....

b) Complete the table below to show the state (solid / liquid / gas) of the sample at different times. (3 marks)

Time (s)	State of sample
25	
60	
65	

c) Explain the shape of the graph between:

i) 5–40 seconds (2 marks)

....

....

ii) 40–55 seconds (2 marks)

....

....

iii) 55–70 seconds (2 marks)

....

....

Score /10

For more help on this topic see KS3 Science Revision Guide pages 42–43.

A Choose just one answer: a, b, c or d.

1 What are the horizontal rows called in the Periodic Table? (1 mark)
   a) rows ○ ✓
   b) groups ○
   c) columns ○
   d) periods ○ ✓

2 What are the vertical columns called in the Periodic Table? (1 mark)
   a) groups ○
   b) rows ○ ✓
   c) columns ○
   d) atoms ○

3 How does the reactivity of Group 1 metals change as you go down the group? (1 mark)
   a) increases ○
   b) decreases ○ ✓
   c) increases then decreases ○
   d) stays the same ○

4 How does the reactivity of the elements in Group 7 change as you go down the group? (1 mark)
   a) increases ○
   b) decreases ○ ✓
   c) increases then decreases ○
   d) stays the same ○

5 The majority of elements in the Periodic Table are: (1 mark)
   a) non-metals ○
   b) radioactive ○
   c) metals ○
   d) gases ○

Score /5

B Answer each question.

1 How are elements arranged in the Periodic Table? (1 mark)

metal / non-metal

2 Who constructed the first version of the modern Periodic Table? (1 mark)

3 Which metal is liquid at room temperature? (1 mark)

4 What is a non-metal oxide? (1 mark)

Score /4

1 Read each description and write either 'metal' or 'non-metal' alongside. (4 marks)

Malleable ....

Found in Group 2 ....

Oxide forms an acid when dissolved in water ....

Might be a gas at room temperature ....

2 The graph below shows the densities of different solid elements.

a) Which two elements shown in the graph are most likely to be metals? (2 marks)

....

b) Explain your answer to part **a)**. (2 marks)

....

....

c) Density is not a very accurate way to determine if an element is a metal or a non-metal. In order to obtain more accurate results, oxides of the elements were produced.

Explain how the oxides could be used to check which elements are metals and which are non-metals. (4 marks)

....

....

....

....

Score /12

For more help on this topic see KS3 Science Revision Guide pages 44–45.

A Choose just one answer: a, b, c or d.

1 What term refers to how readily an element will react? (1 mark)
a) interactivity ◯
b) inactivity ◯
c) reactioness ◯
d) reactivity ◯

2 What is the name of the rocks in which metals are found? (1 mark)
a) crystals ◯
b) metalite ◯
c) ores ◯
d) prions ◯

3 Which of the following metals is usually dug out of the ground as a pure metal? (1 mark)
a) aluminium ◯
b) gold ◯
c) iron ◯
d) zinc ◯

4 Which type of reactions can be used to extract metals from the rocks in which they are found? (1 mark)
a) replacement reactions ◯
b) oxidation reactions ◯
c) displacement reactions ◯
d) combustion reactions ◯

5 Which non-metal is often used in the process to extract metals from the rocks they are found in? (1 mark)
a) nitrogen ◯
b) silicon ◯
c) argon ◯
d) carbon ◯

Score /5

B Answer each question.

1 What are ceramics made from? (1 mark)

....

2 Give two uses of ceramics. (2 marks)

....

....

3 What are composites made from? (1 mark)

....

4 Give two examples of composites. (2 marks)

....

....

5 Give an example of a polymer. (1 mark)

....

Score /7

1 The diagrams below show ethene and polyethene.

Ethene

Polyethene

a) In the diagrams above, which is the monomer and which is the polymer? (1 mark)

Monomer ....

Polymer ....

b) Explain your answers to part **a)**. (1 mark)

....

....

c) State one use of polymers. (1 mark)

....

2 a) Put the following metals in order of reactivity, starting with the most reactive. (3 marks)

zinc   sodium   aluminium   iron   lead

....

....

b) Which of the metals in part **a)** can be extracted from their ores using zinc? (2 marks)

....

c) Explain your answer to part **b)**. (2 marks)

....

....

d) Explain why a metal would not be required to extract platinum from its ore. (2 marks)

....

....

....

Score **/12**

For more help on this topic see KS3 Science Revision Guide pages 46–47.

A Choose just one answer: a, b, c or d.

1 What are sedimentary rocks formed from? **(1 mark)**
- a) lava ⃝
- b) gases ⃝
- c) sediments ⃝
- d) diamonds ⃝

2 What type of rock is basalt? **(1 mark)**
- a) sedimentary ⃝
- b) metamorphic ⃝
- c) igneous ⃝
- d) fossilised ⃝

3 What type of rock is sandstone? **(1 mark)**
- a) sedimentary ⃝
- b) igneous ⃝
- c) metamorphic ⃝
- d) polymer ⃝

4 What is the outer layer of the Earth called? **(1 mark)**
- a) inner core ⃝
- b) outer core ⃝
- c) mantle ⃝
- d) crust ⃝

5 In what form is carbon most commonly found in the Earth's atmosphere? **(1 mark)**
- a) carbon monoxide ⃝
- b) carbon oxide ⃝
- c) carbon dioxide ⃝
- d) carbon trioxide ⃝

Score   /5

B Answer each question.

1 State two of the main elements that the Earth is made up of. **(2 marks)**

....

2 Which part of the Earth is the hottest? **(1 mark)**

....

3 Which type of rock contains fossils? **(1 mark)**

....

4 How are metamorphic rocks formed? **(3 marks)**

....

....

....

Score   /7

Answer all parts of the question. Use a separate sheet of paper if necessary.

1 The graph below shows how the concentration of atmospheric carbon dioxide has changed over time.

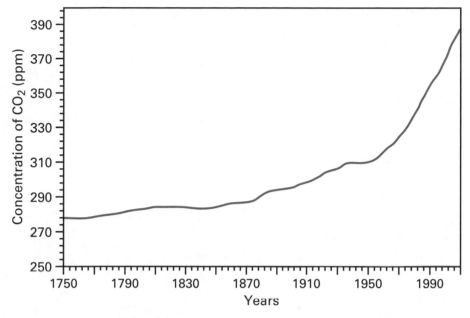

a) Describe the trend of the graph. (3 marks)

....

....

....

b) What problem is the increase in carbon dioxide concentration believed to be causing? (1 mark)

....

c) Explain why the following are increasing the concentration of carbon dioxide in the atmosphere:

   i) using coal to produce electricity in power stations (2 marks)

   ....

   ....

   ii) cutting down trees and then burning them (3 marks)

   ....

   ....

   ....

   ....

Score   /9

For more help on this topic see KS3 Science Revision Guide pages 48–49.

MODULE 21

CALCULATION OF FUEL USES AND COSTS IN THE DOMESTIC CONTEXT

A Choose just one answer: a, b, c or d.

1 Which unit is used to measure energy in food? (1 mark)
a) watts ○
b) ohms ○
c) amps ○
d) kilojoules ○

2 Which term is used for energy transferred over time? (1 mark)
a) current ○
b) potential difference ○
c) power ○
d) resistance ○

3 Which unit is used to measure power? (1 mark)
a) volts ○
b) amps ○
c) joules ○
d) watts ○

4 Which unit is normally used when calculating household energy bills? (1 mark)
a) kWh ○
b) Ws ○
c) Jmin ○
d) sAmp ○

5 Which of the following is not a fossil fuel? (1 mark)
a) coal ○
b) biomass ○
c) gas ○
d) oil ○

Score /5

B Answer each question.

1 How can you find out how much energy is in a food? (1 mark)

....

2 What type of fuels are formed from the remains of dead organisms? (1 mark)

....

3 Why are some energy sources known as 'non-renewable'? (1 mark)

....

4 Why are some energy sources known as 'renewable'? (1 mark)

....

Score /4

STRUCTURE AND FUNCTION OF LIVING ORGANISMS

Module 1: Cells and organisation (pages 4–5)

A

1. d 2. d 3. b 4. a 5. d

B

1. Cells
2. Diffusion
3. Cytoplasm
4. Tissue
5. Organ systems
6. Multicellular

C

1. A – Chloroplast
   B – Cell wall
   C – Vacuole
   D – Mitochondria
   E – Nucleus
2. Cell; Tissue; Organ; Organ system
   (2 marks if fully correct; 1 mark if two correct)

Module 2: The skeletal and muscular systems (pages 6–7)

A

1. a 2. c 3. b 4. c 5. c

B

1. It pulls *(1 mark)* and shortens *(1 mark)*
2. It lengthens
3. A pair of muscles that work opposite to each other
4. Relaxing

C

1.

Function	Description
Movement	Provides an anchor point for muscles
Protection	Prevents the organs from being damaged
Production of blood cells	Cells are produced in the bone marrow
Support	Provides structure

(1 mark for each correct row)

2. a)

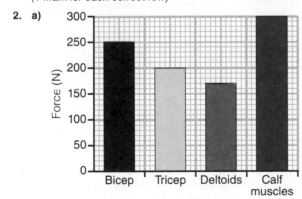

Bars plotted correctly *(2 marks)*
Correct horizontal axis label *(1 mark)*
Correct vertical axis label *(1 mark)*

  b) Repeating the investigation and calculating a mean result will improve the repeatability of the results.

Module 3: Nutrition and digestion (pages 8–9)

A

1. b 2. c 3. c 4. a 5. d

B

1. They release energy / in respiration / excess carbohydrates are converted to fat to store *(any two for 2 marks)*
2. Enzymes are biological catalysts *(1 mark)* which speed up the breakdown of food in the digestive system *(1 mark)*

3. Bacteria help digestion *(1 mark)* and produce vitamins (e.g. vitamin K) *(1 mark)*
4. Obesity
5. Produces bile which neutralises stomach acid *(1 mark)* and helps the digestion of fats *(1 mark)*

C

1. a) Diet B *(1 mark)*. It has a much higher proportion of lipids than the healthy, balanced diet *(1 mark)*. Taking in too many lipids leads to obesity *(1 mark)*.
   b) Nutrient deficiencies (e.g. rickets, anaemia, etc.)

2.

	Organ name	Function
A	Oesophagus	The foods travels down the oesophagus from the mouth to the stomach
B	Small intestine	Chemical digestion / absorption of products of digestion
C	Large intestine	Water is absorbed in the large intestine / transports undigested food from the small intestine to the rectum

(1 mark for each correct row)

Module 4: Gas exchange system in humans (pages 10–11)

A

1. d 2. a 3. c 4. b 5. b

B

1. The gas exchange system supplies oxygen / for respiration / and removes carbon dioxide / produced in respiration *(any three for 3 marks)*.
2. Pressure is increased *(1 mark)*, forcing air out of the lungs *(1 mark)*.
3. The bronchioles constrict *(1 mark)*, preventing air getting in and out of the lungs *(1 mark)*.
4. Drugs *(1 mark)* given using an inhaler *(1 mark)*

C

1.

	Organ name	Function
A	Mouth	Allows air into and out of the body
B	Trachea	Carries air from the mouth and nose to the bronchi
C	Alveoli / lung	Where oxygen diffuses into the blood and carbon dioxide diffuses out of the blood

(1 mark for each correct row)

2. a) Cars create air pollution *(1 mark)*; air pollution can cause asthma *(1 mark)*
   b) Yes *(1 mark)*. Increases in asthma could be caused by other factors *(1 mark)*; no proof that car pollution has caused the increase *(1 mark)*

Module 5: Human reproduction (pages 12–13)

A

1. c 2. a 3. c 4. d 5. c

B

1. Alcohol and drugs in the mother's blood *(1 mark)* are able to pass into the baby's blood / via the placenta *(1 mark)*.
2. In the oviducts *(1 mark)* in the female's body *(1 mark)*
3. In the testes *(1 mark)* in the male's body *(1 mark)*
4. In the uterus *(1 mark)* in the female's body *(1 mark)*

C

1. The zygote divides to form an embryo then moves down the oviduct to the uterus where it implants in the endometrium. – 4

After nine months the mother will go into labour. Muscle contractions push the foetus out of the vagina and the baby is born. – 6

Once a month, an egg is released from the ovaries and the lining of the uterus (the endometrium) thickens in preparation for implantation. – 1

If fertilisation does take place the sperm will meet the egg in the oviduct. – 2

The embryo develops into a foetus and is provided with nutrients from the mother's blood by the placenta. – 5

The sperm and egg fuse to form a zygote. – 3

(4 marks if fully correct; otherwise 1 mark for each correct answer)

2. The egg contains stored food *(1 mark)* for the developing embryo *(1 mark)*.

3. The lining of the uterus (the endometrium) breaks down and is released in menstruation *(1 mark)*. This cannot occur when the woman is pregnant as the developing embryo or foetus is implanted in the lining of the uterus *(1 mark)*.

Module 6: Plant reproduction (pages 14–15)

A

1. c **2.** d **3.** b **4.** a **5.** a

B

1. Allows the male gamete *(1 mark)* to reach the female gamete in the flower *(1 mark)*

2. Insect-pollinated flowers have bright coloured petals and nectar to attract insects. Wind-pollinated flowers have dull coloured petals and no nectar. They do not need to attract insects. *(1 mark for each correct difference up to a maximum of 3)*

3. They attach to the fur of animals *(1 mark)*, which can then carry the seeds away *(1 mark)*.

C

1. a) i) Dispersal by animals
   ii) Plant A is only found in the area where the animals live.
   b) i) Dispersal by wind
   ii) Plant B is found in areas where animals do not live / seeds can cross mountainous region.
   c) Wind pollination *(1 mark)*. Pollen grains can be carried by the wind across the mountainous region *(1 mark)*.

Module 7: Health (pages 16–17)

A

1. c **2.** d **3.** d **4.** a **5.** a

B

1. Injecting heroin with shared / dirty needles
2. When people sniff aerosols or glue
3. Nicotine
4. Tar
5. LSD causes hallucinations *(1 mark)* so users might behave in dangerous ways without realising *(1 mark)*.

C

1. a) The number of deaths increases from 1970 to 1980 *(1 mark)*; the number of deaths levels off between 1980 and 1990 *(1 mark)*; the number of deaths increases again between 1990 and 2005 *(1 mark)*.
   b) Yes – as people wouldn't be drinking *(1 mark)*, so there would be less liver disease caused by excess drinking *(1 mark)* or: No – people would still drink illegally *(1 mark)*, so there would still be high liver disease due to excess drinking *(1 mark)*.

2. Alcohol affects reaction times / decision making / co-ordination *(2 marks for any two of these effects)*, therefore it increases the chance of someone being involved in an accident whilst driving *(1 mark)*.

BIOLOGICAL PROCESSES

Module 8: Photosynthesis (pages 18–19)

A

1. b **2.** d **3.** a **4.** c **5.** d

B

1. Carbon dioxide *(1 mark)*; water *(1 mark)*

2. Glucose *(1 mark)*; oxygen *(1 mark)*
3. Light
4. Reduce photosynthesis / less carbon dioxide absorbed from the atmosphere / carbon dioxide levels in atmosphere increase / leads to global warming and climate change *(1 mark for each up to a maximum of 3)*

C

1. a)

Sample	Photosynthesis occurs	Explanation
A	No	No light energy present
B	No	No carbon dioxide present
C	No	No carbon dioxide present and no chloroplasts for photosynthesis to occur
D	Yes	All conditions for photosynthesis to occur are present

(1 mark for each correct row)

   b) Oxygen *(1 mark)*. Oxygen is produced by photosynthesis and the faster the rate of photosynthesis, the faster the rate of oxygen production *(1 mark)*.

Module 9: Cellular respiration (pages 20–21)

A

1. c **2.** d **3.** b **4.** a **5.** a

B

1. Oxygen / glucose
2. Carbon dioxide / water / energy
3. Glucose
4. Alcohol (ethanol) / carbon dioxide
5. Lactic acid

C

1. a) No air / oxygen should enter the apparatus.
   b) Carbon dioxide
   c) No *(1 mark)*. Human cells do not produce carbon dioxide when they anaerobically respire *(1 mark)*, so there would be no gas to collect *(1 mark)*.
   d) Woodlice are animals so it could be considered cruel to experiment on them.

Module 10: Relationships in an ecosystem (pages 22–23)

A

1. d **2.** c **3.** a **4.** b **5.** a

B

1. To kill pests / that eat and damage crops / which reduces the number of crops that can be produced *(1 mark for each up to a maximum of 2)*.

2. The primary consumers are eaten by secondary consumers and pass on the toxic materials to them / could also be passed to a tertiary consumer / the toxic material accumulates in the food chain *(1 mark for each up to a maximum of 2)*.

3. Pesticides may kill pollinating insects / so food or crop plants cannot reproduce / which reduces the amount of food that can be produced *(1 mark for each)*.

C

1. a)

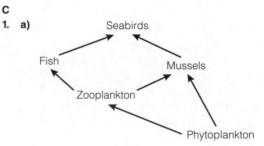

(1 mark for each correct part of food web)

b) i) Phytoplankton
   ii) Zooplankton / Mussels
   iii) Mussels / fish / seabirds

Module 11: Inheritance, chromosomes, DNA and genes (pages 24–25)

A

1. d **2.** b **3.** d **4.** a **5.** c

B

1. Gene banks store embryos or seeds of organisms *(1 mark)*. If an organism becomes extinct the embryos or seeds can be used *(1 mark)* to reintroduce the organism *(1 mark)*.
2. Genetic information is passed *(1 mark)* from one generation to the next *(1 mark)*.
3. Discontinuous variation *(1 mark)*. Eye colours are divided into categories *(1 mark)*.

C

1.

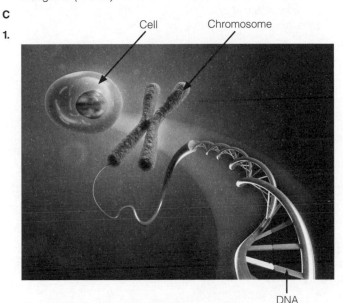

Cell   Chromosome

DNA

(2 marks if fully correct; 1 mark if one correct)

2. a) A and D are the same species.
   b) A and D are able to breed together to produce fertile offspring / therefore they are the same species *(1 mark)*. None of the other populations are able to breed together to produce fertile offspring *(1 mark)*.

CHEMISTRY
Module 12: The particulate nature of matter (pages 26–27)

A

1. a **2.** b **3.** a **4.** d **5.** b

B

1. By heating the gas *(1 mark)* or by decreasing the volume of the container the gas is in *(1 mark)*.
2. By cooling the gas *(1 mark)* or increasing the volume of the container the gas is in *(1 mark)*.
3. It gains kinetic energy *(1 mark)* and it will move faster *(1 mark)*.

C

1. a) i) Solids have a fixed shape and volume *(1 mark)* so will provide support to the house *(1 mark)*.
   ii) The volume and shape of a gas are not fixed *(1 mark)*, so the gas can be used to fill the balloon *(1 mark)*.
   iii) The shape of liquids is not fixed so people can swim *(1 mark)*. Liquids have a fixed volume so they will stay in the pool *(1 mark)*.
   b) The gas particles / gain kinetic energy. They collide with the side of the balloon / more frequently / the pressure of the gas in the balloon increases. *(1 mark for each up to a maximum of 3)*
   c) Liquid *(1 mark)*. Liquid is the only state of matter which is difficult to compress *(1 mark)* and can flow *(1 mark)*.

Module 13: Expansion and contraction (pages 28–29)

A

1. b **2.** a **3.** d **4.** a **5.** a

B

1. The particles lose energy / and this causes the particles to move slower / the particles are no longer able to overcome the attractive forces / the particles are no longer able to move relative to each other / the shape of the liquid becomes fixed and it is now a solid *(1 mark for each up to a maximum of 3)*
2. The particles gain energy / and this causes the particles to vibrate faster / the particles are able to break some of the attractive forces / the particles are now able to move relative to each other *(1 mark for each up to a maximum of 3)*

C

1. a) i) Zinc
       ii) Iron
   b) Heating
   c) Liquid mercury or alcohol in a thermometer / when the temperature rises / the liquid expands and rises up the thermometer a certain distance / showing the temperature / when the liquid cools it contracts / the liquid moves down the tube a certain distance / showing the new temperature *(1 mark for each up to a maximum of 4)*

Module 14: Atoms, elements and compounds (pages 30–31)

A

1. c **2.** d **3.** d **4.** c **5.** a

B

1. A substance that contains more than one different type of element bonded together.
2. It remains the same
3. It remains the same
4. 5g

C

1. a)

Particle	Relative mass	Charge
Proton	1	Positive
Neutron	1	Neutral
Electron	$\dfrac{1}{1800}$	Negative

(1 mark for each correct row)

   b) A – proton *(1 mark)*, B – neutron *(1 mark)*, C – electron *(1 mark)*
2. No *(1 mark)*. Mass is always conserved in a reaction *(1 mark)*. The mass of the wood has been transferred to the ash and the smoke produced *(1 mark)*.

Module 15: Pure and impure substances (pages 32–33)

A

1. d **2.** b **3.** b **4.** a **5.** c

B

1. Lower the melting point *(1 mark)* and raise the boiling point *(1 mark)*
2. Evaporation; distillation; chromatography; filtration *(any two for 2 marks)*
3. The substance which dissolves
4. The liquid the solute dissolves in
5. The liquid produced when a solute dissolves in a solvent

C

1. a) A – Copper sulfate *(1 mark)*; B – Sand *(1 mark)*, C – Salt *(1 mark)*
   b) Diffusion
2. a) Chromatography
   b) B, A, C *(2 marks if fully correct; 1 mark if one correct)*
   c) The more soluble the dye, the further it moves up the chromatography paper.

Module 16: Chemical reactions (pages 34–35)

A

1. c **2.** a **3.** b **4.** c **5.** a

B

1. Oxidation
2. The strength of an acid or alkali
3. pH 7
4. An indicator
5. They are corrosive
6. Safety goggles

C

1. **a)** weak acid + weak alkali → **neutral**
    strong alkali + **weak acid** → weak alkali
    strong acid + **strong alkali** → neutral
    weak alkali + strong acid → weak acid
    (2 marks for each correct pair of answers)

    b) An indicator, e.g. universal indicator; phenolphthalein

2. A substance is burned in air – combustion
    A substance breaks down when heated – thermal decomposition
    A reaction with oxygen to form an oxide – oxidation
    An element replaces a more reactive element in a compound – displacement
    (3 marks if fully correct; 2 marks if two correct; 1 mark if one correct)

3. A catalyst *(1 mark)*. It speeded up the reaction *(1 mark)*.

Module 17: Energetics (pages 36–37)

A

1. a **2.** b **3.** c **4.** b **5.** a

B

1. The temperature stays the same
2. Endothermic reaction
3. Exothermic reaction
4. The temperature would decrease as it cooled *(1 mark)*. As the gas was condensing, the temperature would remain constant *(1 mark)* but then the temperature would carry on decreasing *(1 mark)*.

C

1. **a)** 40 seconds

    b)

Time (s)	State of sample
25	Liquid
60	Gas
65	Gas

    (1 mark for each correct row)

    c) i) The liquid is being heated / the particles are gaining energy / so the temperature is increasing *(1 mark for each up to a maximum of 2)*

    ii) As the liquid is changing into a gas *(1 mark)* the temperature remains constant *(1 mark)*

    iii) The gas is being heated / the particles are gaining energy / so the temperature is increasing *(1 mark for each up to a maximum of 2)*

Module 18: The Periodic Table (pages 38–39)

A

1. d **2.** a **3.** a **4.** b **5.** c

B

1. By relative atomic mass
2. Dmitri Mendeleev
3. Mercury
4. A compound formed by the oxidation of a non-metal element

C

1. Malleable – metal; Found in Group 2 – metal; Oxide forms an acid when dissolved in water – non-metal; Might be a gas at room temperature – non-metal *(1 mark for each)*

2. **a)** B *(1 mark)* and E *(1 mark)*

    b) They have the highest densities *(1 mark)*. Generally metals are more dense than non-metals *(1 mark)*.

    c) Dissolve both oxides in water *(1 mark)* and test the pH of each *(1 mark)*. Metal oxides would form an alkaline solution *(1 mark)* and non-metal oxides would form an acidic solution *(1 mark)*.

Module 19: Materials (pages 40–41)

A

1. d **2.** c **3.** b **4.** c **5.** d

B

1. Crystalline compounds
2. Any suitable answers, e.g. tiles; pots; plates *(1 mark for each up to a maximum of 2)*
3. Composites are made from two or more different materials.
4. Any suitable answers, e.g. concrete; carbon fibre *(1 mark for each up to a maximum of 2)*
5. Any suitable answer, e.g. PVC; polyethene

C

1. **a)** Monomer – ethene; polymer – polyethene *(1 mark for both)*

    b) Polyethene is made up of many monomers but ethene is a single unit.

    c) Any suitable answer, e.g. packaging; fibres

2. **a)** Sodium, aluminium, zinc, iron, lead *(3 marks if fully correct; 2 marks if one error; 1 mark if two errors)*

    b) Iron *(1 mark)* and lead *(1 mark)*

    c) Iron and lead are less reactive than zinc *(1 mark)* so they can be extracted with a displacement reaction from their ores *(1 mark)*.

    d) Platinum is very unreactive / so does not form an ore / it can be dug straight out of the ground *(1 mark for each up to a maximum of 2)*

Module 20: Earth and atmosphere (pages 42–43)

A

1. c **2.** c **3.** a **4.** d **5.** c

B

1. Any two from: iron; oxygen; silicon; magnesium; sulfur; nickel; calcium; aluminium
2. The inner core
3. Sedimentary
4. Igneous or sedimentary rocks / are heated at high pressure / beneath the surface of the Earth / and become different types of rock *(1 mark for each up to a maximum of 3)*

C

1. **a)** Carbon dioxide concentration increases slowly over time *(1 mark)* until 1950 *(1 mark)*, then the concentration increases rapidly over time *(1 mark)*

    b) Global warming / climate change

    c) i) Burning coal *(1 mark)* releases carbon dioxide into the atmosphere *(1 mark)*.

    ii) Cutting down trees reduces the amount of photosynthesis which occurs *(1 mark)* so less carbon dioxide is absorbed from the atmosphere *(1 mark)*. Burning trees also releases carbon dioxide into the atmosphere *(1 mark)*.

ENERGY

Module 21: Calculation of fuel uses and costs in the domestic context (pages 44–45)

A

1. d **2.** c **3.** d **4.** a **5.** b

B

1. Look at the label
2. Fossil fuels
3. They will eventually run out
4. They will not eventually run out

C

1. a)

Renewable	Non-renewable
Wind	Coal
Tidal	Oil
Geothermal	Gas

(1 mark for each correct row)

b) The Sun doesn't always shine *(1 mark)*. Solar panels wouldn't always generate electricity / wouldn't generate electricity at night when it is dark *(1 mark)*.

2. a) Family A: October–December *(1 mark)*
Family B: January–March *(1 mark)*

b) Family A: £400 *(1 mark)*
Family B: £350 *(1 mark)*

c) Family A *(1 mark)*; £1050 *(1 mark)*

Module 22: Energy changes and transfers (pages 46–47)

A

1. c **2.** a **3.** c **4.** c **5.** d

B

1. Cavity wall insulation / double glazing / loft insulation *(1 mark for each up to a maximum of 2)*

2. No *(1 mark)*. It can only be transferred *(1 mark)*.

3. Chemical to electrical

4. Radiation

C

1. a) From A to B

b) Conduction

c) They would become the same temperature

d) Yes *(1 mark)*. The transfer of heat would be slower / as the insulator reduces heat loss / the increase in the temperature of B would be slower / it would take longer for both objects to become the same temperature *(1 mark for each up to a maximum of 2)*

Module 23: Changes in systems (pages 48–49)

A

1. c **2.** b **3.** a **4.** c **5.** b

B

1. 13kJ/g

2. Chemical energy is converted to light energy *(1 mark)*, thermal energy *(1 mark)* and chemical energy in products *(1 mark)*.

3. a) Elastic potential

b) Converted to kinetic energy

c) Pulled back

C

1. a) i) When it is moving

ii) When the car is moving it has kinetic energy *(1 mark)*. When it has stopped moving it does not have kinetic energy *(1 mark)*.

b) Chemical *(1 mark)* to kinetic *(1 mark)*

c) Chemical *(1 mark)* to electrical / to sound / to light *(1 mark)*

d) Kinetic energy is transferred / to heat / and sound / by the brakes *(1 mark for each up to a maximum of 2)*. This energy is lost to the environment *(1 mark)*.

MOTION AND FORCES
Module 24: Describing motion (pages 50–51)

A

1. a **2.** b **3.** a **4.** a **5.** c

B

1. 2m/s

2. 33.3m/s

3. 66.7m/s

4. 1050m

5. 700 seconds

C

1. a) 10–20 seconds

b) This is when the line is the steepest.

c) 20–25 seconds

d) This is when the line is flat.

e) The graph would rise steeply *(1 mark)*, then continue rising at a less steep angle *(1 mark)*, before flattening out *(1 mark)*.

Module 25: Forces, motion and balanced forces (pages 52–53)

A

1. d **2.** c **3.** d **4.** b **5.** a

B

1. Balanced

2. Remain stationary

3. Any two from: gravity, magnetism, static electricity

4. The force needed to extend or compress a spring *(1 mark)* is proportional to the distance that it is being extended or compressed *(1 mark)*.

5. The pull force on the object will be greater than the opposing force of the object.

C

1. a) Speeding up

b) The forward force is greater than the opposing force.

c) Air resistance / friction

d) No

e) The opposing force has increased since the spoiler was fitted.

f) 7500N

Module 26: Pressure in fluids (pages 54–55)

A

1. a **2.** b **3.** c **4.** a **5.** d

B

1. a) It decreases

b) Higher in the atmosphere there are fewer air particles *(1 mark)*, therefore less force pushing down on you *(1 mark)*.

2. a) It increases

b) Deeper beneath the water there are more water particles above you *(1 mark)*, therefore more force pushing down on you *(1 mark)*.

C

1. a)

Floats	Sinks
A and C	B and D

(2 marks if fully correct; 1 mark if any two answers correct)

b) Materials that are more dense than water will sink *(1 mark)*. Materials that are less dense than water will float *(1 mark)*.

c) Upthrust

2. A submarine dives underwater *(1 mark)* and the deeper the submarine travels, the greater the pressure *(1 mark)*. The submarine has to be built to withstand the high pressures *(1 mark)* to prevent it being damaged or crushed *(1 mark)*.

WAVES
Module 27: Observed waves, sound waves and energy and waves (pages 56–57)

A

1. c **2.** a **3.** d **4.** c **5.** b

B

1. A low-pitched sound

2. The range of frequencies *(1 mark)* an organism can hear *(1 mark)*.

3. Sound that has a higher frequency than the auditory range of humans / above 20 kHZ

4. Any suitable answers, e.g. cleaning; physiotherapy

C

1. **a)** Sound travels faster in non-metals than in metals
   b) Not very reliable *(1 mark)* due to the small sample size *(1 mark)*
   c) Investigate the speed of sound in other metals and non-metals *(1 mark)* to see if sound is always faster in non-metals *(1 mark)*
2. Yes *(1 mark)*. Space is a vacuum *(1 mark)* and sound cannot travel in a vacuum *(1 mark)*. Therefore when the ship exploded no sound would be heard *(1 mark)*.

Module 28: Light waves (pages 58–59)

A

1. a   2. c   3. d   4. a   5. b

B

1. The light splits *(1 mark)* into the colours of the spectrum *(1 mark)*.
2. Angle of reflection
3. No light is reflected
4. All colours of light are reflected

C

1. **a)** A – specular reflection *(1 mark)*
   B – Diffuse scattering *(1 mark)*
   b) A
   c) Light passes through transparent materials
2.

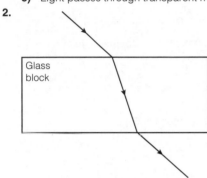

(2 marks if fully correct; 1 mark if one error in the rays drawn)

ELECTRICITY AND ELECTROMAGNETISM
Module 29: Current electricity (pages 60–61)

A

1. c   2. b   3. d   4. a   5. b

B

1. Causes charge to flow
2. Ammeter
3. Voltmeter
4. Insulator

C

1. From top to bottom: bulb; switch; ammeter; voltmeter; resistor
   (3 marks if fully correct; 2 marks if one error; 1 mark if two correct)
2. **a)** Series circuit *(1 mark)*. Current is constant *(1 mark)* in all parts of a series circuit *(1 mark)*
   b) Parallel circuit *(1 mark)*. Current is not constant / in all parts of a parallel circuit / divides down branches *(1 mark for each up to a maximum of 2)*

Module 30: Static electricity and magnetism (pages 62–63)

A

1. d   2. d   3. a   4. b   5. a

B

1. When the same poles (N and N or S and S) *(1 mark)* are brought together *(1 mark)*
2. When opposite poles *(1 mark)* are brought together *(1 mark)*
3. A wire *(1 mark)* is wrapped around an iron core *(1 mark)* and a current is passed through the wire *(1 mark)*.

C

1. **a)**

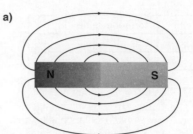

   b)

Material	Attracted
Steel	✓
Cotton	✗
Nickel	✓
Paper	✗

(1 mark for each pair of correct rows)

   c) Compasses work by pointing towards magnetic north *(1 mark)*. The compass could point towards the bar magnet instead *(1 mark)*, meaning you would not be sure which way is north *(1 mark)*.
2. Suspend a coil of wire *(1 mark)* between the opposite poles of two magnets *(1 mark)*. When an electric current is passed through the wire *(1 mark)*, the wire will experience a force and turn *(1 mark)*.

MATTER AND SPACE PHYSICS
Module 31: Physical changes (pages 64–65)

A

1. a   2. d   3. d   4. b   5. b

B

1. Chemical change
2. Change of state
3. Brownian motion
4. Liquid *(1 mark)* and solid *(1 mark)*
5. Gas

C

1. **a)** No
   b) This is dissolving *(1 mark)*. The solute has dissolved in the solvent to form a solution *(1 mark)*
   c) Evaporation
   d) **i)** 20g
   ii) 15g
   e) The gas could be cooled *(1 mark)*, so the gas would condense to form a liquid *(1 mark)*.

Module 32: Particle model and energy in matter (pages 66–67)

A

1. a   2. b   3. d   4. d   5. c

B

1. Burning the fuel
2. Ice is a solid / which is less dense than liquid water / because the bonds between the molecules in the ice / force them apart / so the molecules are further apart in ice than in liquid water
   (1 mark for each up to a maximum of 3)
3. Respiration

C

1. Particles very close together – 1
   Particles very far apart – 3
   Particles quite close together – 2
   (2 marks if fully correct; 1 mark if one correct)
2. **a)** Ice
   b) The ice will float

c) The **particles** in the **ice** will gain **energy**. This will cause the **particles** to move faster, breaking the **bonds** between the particles. The **ice** will melt, forming **liquid** water.
(3 marks if fully correct; 2 marks if up to two errors; 1 mark for three correct answers)

Module 33: Space physics (pages 68–69)

A
1. c **2.** d **3.** a **4.** d **5.** c

B
1. 15 000N
2. 576N
3. During the summer the Earth is tilted towards the Sun / so receives more of the Sun's energy. During the winter the Earth is tilted away from the Sun / so receives less of the Sun's energy / so when it is summer in the northern hemisphere it is winter in the southern hemisphere, and vice versa.
   (1 mark for each up to a maximum of 3)
4. Weight

C
1. a) Jupiter; Neptune; Earth; Venus; Mars
   b) i) Mars
      ii) Jupiter
   c) Gravity is caused by mass *(1 mark)*. The planet with the strongest gravity would have the largest mass *(1 mark)* and vice versa.

MIXED TEST-STYLE QUESTIONS

1.
Element	Compound
Ba, Ru, O_2	HCl, NaF, CO_2

(1 mark for each correct pair of answers)

2. a) A – Crust
      B – Mantle
      C – Outer core
      D – Inner core
      (1 mark for each)
   b)
Solid	Liquid
Crust Mantle Inner core	Outer core

   (1 mark for each correct pair of answers)

3.
Distance	Most appropriate unit
The diameter of a galaxy	light years
The distance between two cities	km
The width of a school sports field	metres
The length of an insect	mm

(1 mark for each correct pair of answers)

4. A – Respiration; B – Photosynthesis; C – Decomposition
   (1 mark for each)
5. A – Cell membrane; B – Cytoplasm; C – Chloroplast; D – Vacuole
   (1 mark for each)
6. a) A – Solid; B – Liquid; C – Gas
      (1 mark for each)
   b)
State of matter	Shape	Volume	Compressible
Gas	No fixed shape	No fixed volume	Easily compressed
Solid	Fixed shape	Fixed volume	Difficult to compress
Liquid	No fixed shape	Fixed volume	Difficult to compress

(3 marks if fully correct; 2 marks if up to two errors; 1 mark if three answers correct)

7. a) A series circuit
   b) A bulb
   c) The circuit is broken *(1 mark)*, so the current can't reach the bulb *(1 mark)*.

8. a) Photosynthesis
   b) It would die
   c) i) It would die (as it would be unable to feed on the primary producer)
      ii) It would die (as it would be unable to feed on the primary consumer)

9.
Metal	Description
Zinc	Can be extracted from its ore using carbon and is more reactive than iron
Silver	Doesn't form an ore
Copper	Less reactive than lead but more reactive than silver
Potassium	Most reactive metal in this table

(3 marks if fully correct; 2 marks if one error; 1 mark if two answers correct)

10. a)
	Oxygen	Carbon dioxide
Inhaled	High	Low
Exhaled	Low	High

   (1 mark for each correct row)
   b) Oxygen is taken in and used in respiration *(1 mark)*. Carbon dioxide is produced in respiration and needs to be released *(1 mark)*.

11.
Substance moves from the solution into the cell	Substance moves from the cell into the solution	There is no net movement of substance
A D	B	E C

(1 mark for each correct column)

12. a) Thermal energy
    b) Higher *(1 mark)*. The eggs gained energy from the stove *(1 mark)*.
    c) The eggs would lose energy / to the environment / so would cool down / to the same temperature as the rest of the room
       (1 mark for each up to a maximum of 3)

13. a) B *(1 mark)*. B has the highest temperature *(1 mark)*. Pressure in a gas increases as temperature increases *(1 mark)*.
    b) C *(1 mark)*. C has the smallest volume *(1 mark)*. Pressure in a gas increases as volume decreases *(1 mark)*.

14. The vibrations of the diaphragm are converted to an electrical signal – 5
    The singer's vocal chords vibrate – 1
    The sound waves hit the diaphragm of the microphone – 3
    A longitudinal sound wave is produced – 2
    The diaphragm of the microphone vibrates – 4
    The electrical signal is transmitted to the recording device down a wire – 6
    (3 marks if fully correct; 2 marks if four statements in correct order; 1 mark if three statements in correct order)

15. A – Organism
    B – Organ
    C – Tissue
    D – Organ system
    (2 marks if fully correct; 1 mark if one error)

16. a) Carbon, hydrogen and oxygen are all **elements**. As substance A is made up of three different **elements** joined together it is a **compound**.

b) Experiment A: change of state
Experiment B: chemical reaction

c)

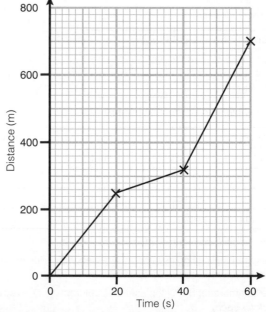

50% Hydrogen
25% Carbon
25% Oxygen

17. Air travels through the bronchi – 3
Air enters the mouth – 1
Air enters the alveoli – 5
Air travels down the trachea – 2
Air enters the bronchioles in the lungs – 4
(2 marks if fully correct; 1 mark if one error)

18. a) Distillation
b) i) It heats the mixture *(1 mark)* to the boiling point of the liquid to be separated from the mixture *(1 mark)*.
ii) It cools the gas *(1 mark)*, forming a pure liquid which can be collected *(1 mark)*.

19. a) A prism
b) red, orange, yellow, green, blue, indigo, violet *(2 marks if fully correct; 1 mark if at least five colours in correct order)*
c) The light refracts when it enters the prism *(1 mark)*. Each colour refracts slightly differently *(1 mark)*.
d) The lens
e) When light hits the retina a chemical reaction occurs / which causes an electrical signal *(1 mark)* to be sent to the brain *(1 mark)*.

20. The external intercostal muscles contract and pull the ribs up and out – I
The pressure in the lungs increases – O
The pressure in the lungs decreases – I
The lungs increase in volume – I
The diaphragm relaxes and raises – O
The diaphragm contracts and flattens – I
The external intercostal muscles relax, the ribs move down and in – O
The lungs decrease in volume – O
(1 mark for each correct pair of answers)

21. a) Solutions / mixtures of solids and liquids
b) So that the liquid can evaporate *(1 mark)* and leave the container *(1 mark)*.
c) By heating the mixtures

22. a)

	Solute / Solvent / Solution	Mass (g)
Salt	Solute	5
Water	Solvent	100
Saltwater	Solution	105

(1 mark for each correct row)

b) No *(1 mark)*. The saltwater is a solution / so it cannot be separated by filtration *(1 mark)* as the solute is dissolved in the solvent *(1 mark)*.

23. A – Penis; B – Testes; C – Vagina; D – Uterus; E – Oviduct;
F – Ovary *(3 marks if fully correct; otherwise 1 mark for each correct pair of answers)*

24. a) Anaerobic respiration
b) Lactic acid
c) Less energy produced / than in aerobic respiration
d) i) It increases vital capacity or volume of the lungs
ii) It uses carbohydrate or fat to release energy
e) Asthma causes the bronchioles to constrict / reduces the amount of oxygen that can get to the lungs *(1 mark)* and reduces the amount of aerobic respiration that can occur *(1 mark)*.

25. a) Molten (liquid) rock or lava or magma *(1 mark)* cools to form solid rock *(1 mark)*.

b) Rock B *(1 mark)*. Intrusive igneous rocks / cool slowly underground / so have large crystals.
(1 mark for each up to a maximum of 2)
c) Rock A *(1 mark)*. Extrusive igneous rocks / cool quickly on the Earth's surface / so have small crystals.
(1 mark for each up to a maximum of 2)

26. a) An electromagnet
b) The same shape as the magnetic field of a bar magnet
c) When a current is passed through the wire
d) By increasing the current *(1 mark)* or increasing the number of turns on the wire *(1 mark)*

27. DDT is sprayed on crops – 1
Small birds feed on many insects – 3
The population of hawks decreases – 8
Insects feed on the crops and take in DDT – 2
Due to DDT, the hawks' eggs are very fragile and break – 7
Hawks feed on many small birds – 5
DDT accumulates in the small birds' bodies – 4
DDT accumulates in the hawks' bodies – 6
(3 marks if fully correct; 2 marks if one error; 1 mark if at least three statements are in the correct order)

28. a) Variation
b) Food
c) They are better adapted *(1 mark)* so are able to eat more nuts *(1 mark)*.
d) Natural selection

29. a) An acid
b) Acids have a low pH *(1 mark)* so are corrosive *(1 mark)*.
c) Any one from: yellow; orange; red
d) Any one from: acid + metal $\Rightarrow$ salt + water; neutralisation

30. a) 16.7m/s
b) 334m (or 333.3m)
c) 120 seconds (or 119.8 seconds)
d) Water resistance
e)

(1 mark for correct axis labels; 1 mark for linear scale; 1 mark for correctly plotting points; 1 mark for connecting points with straight lines with a ruler)

31. a) All dinosaurs have died out / there are no dinosaurs left alive
b) Conservation
c) Gene banks require seeds or embryos / so the organism can be reintroduced if it becomes extinct in the wild *(1 mark)*. There are no dinosaur embryos available to put into a gene bank *(1 mark)*.

1   **a)**   Here is a list of different energy resources.

Place them in the appropriate column of the table.   (3 marks)

wind   **coal**   **oil**   **geothermal**   **tidal**   **gas**

Renewable	Non-renewable

b)   Solar energy is an example of a renewable energy source.

Suggest two problems of trying to provide all of the UK's electricity using solar power.   (2 marks)

....

....

....

2   The table below shows the energy use of two families.

Period	Family A energy use (kWh)	Family B energy use (kWh)
Jan–March	6000	7000
April–June	4000	3000
July–September	3000	2000
October–December	8000	6000

a)   During which period did each family use the most energy?   (2 marks)

Family A: ....

Family B: ....

Each family pays 5p per kWh.

b)   Calculate the highest three-month bill paid by each family.   (2 marks)

Family A: ....

Family B: ....

c)   Which family had the highest bill for the entire year and how much did they pay?   (2 marks)

....

....

Score   **/11**

For more help on this topic see KS3 Science Revision Guide pages 54–55.

A — Choose just one answer: a, b, c or d.

1 How can heat be transferred between two touching objects? (1 mark)
a) refraction ○
b) reflection ○
c) conduction ○
d) superposition ○

2 How can heat be transferred between two objects that are not touching? (1 mark)
a) radiation ○
b) reflection ○
c) polarity ○
d) resistance ○

3 Which term describes a material that is a poor conductor of heat? (1 mark)
a) conductor ○
b) magnetic ○
c) insulator ○
d) convector ○

4 Which form of energy is stored in food? (1 mark)
a) light energy ○
b) nuclear radiation ○
c) chemical energy ○
d) kinetic energy ○

5 Which of the following forms of energy is stored when elastic is stretched? (1 mark)
a) light energy ○
b) nuclear radiation ○
c) sound energy ○
d) elastic potential energy ○

Score /5

B — Answer each question.

1 State two ways of reducing heat loss from a house. (2 marks)

....

....

2 Energy can be destroyed. Is this statement correct?

Explain your answer. (2 marks)

....

....

3 What energy transfer occurs when a battery is plugged into a circuit? (1 mark)

....

....

4 How is the Sun's thermal energy transmitted to Earth? (1 mark)

....

Score /6

Answer all parts of the question. Use a separate sheet of paper if necessary.

1 Florence and Harry investigate the energy transfer between different objects.

A simplified version of their experiment is shown below.

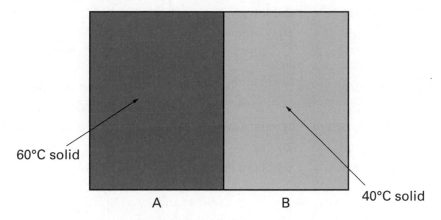

60°C solid

A   B   40°C solid

a) In which direction will the heat energy transfer? (1 mark)

....

b) By what process will the heat energy transfer? (1 mark)

....

c) Predict what would happen if the objects were left for a long period of time. (1 mark)

....

....

d) If object A was wrapped in an insulating material, would the results of the experiment change?

Explain your answer. (3 marks)

....

....

....

....

Score   /6

For more help on this topic see KS3 Science Revision Guide pages 56–57.

A — Choose just one answer: a, b, c or d.

1 How does the total energy at the end of a process compare to the total energy at the start of a process? (1 mark)
a) higher ◯ b) lower ◯
c) the same ◯ d) there is no energy left ◯

2 In an exothermic reaction, how does the energy of the products compare to the energy of the reactants? (1 mark)
a) higher ◯ b) lower ◯
c) the same ◯ d) there is no energy left ◯

3 In an endothermic reaction, how does the energy of the products compare to the energy of the reactants? (1 mark)
a) higher ◯ b) lower ◯
c) the same ◯ d) there is no energy left ◯

4 If a car is slowing down, what happens to its kinetic energy? (1 mark)
a) it stays the same ◯
b) it increases ◯
c) it decreases ◯
d) it increases then decreases ◯

5 When an object is raised above the surface of the Earth, what happens to its gravitational potential energy? (1 mark)
a) it stays the same ◯
b) it increases ◯
c) it decreases ◯
d) it increases then decreases ◯

Score /5

B — Answer all parts of each question.

1 Complete the following equation to show the energy after a chemical reaction. (1 mark)

$$\text{reactants} \rightarrow \text{products}$$
$$13kJ/g \rightarrow$$

2 Describe what happens to the chemical energy stored in a fuel when it is burned. (3 marks)

....

....

....

3 a) What form of energy is stored in a catapult when the elastic is pulled back? (1 mark)

....

b) What happens to this energy when the catapult is released? (1 mark)

....

c) When does the catapult have the highest energy – when it is being pulled back or after it has been released? (1 mark)

....

Score /7

Answer all parts of the question. Use a separate sheet of paper if necessary.

1 A car drives along a long straight road and comes to a stop.

a) i) When does the car have the most energy – when it is moving or when it has stopped? (1 mark)

....

ii) Explain your answer. (2 marks)

....

....

b) When the car is travelling, the engine is burning petrol in order to move the car.

What energy transfer is occurring here? (2 marks)

....

....

c) The driver turns on the car headlights and the radio.

What new energy transfers are occurring? (2 marks)

....

....

....

d) When the car brakes, its total energy decreases.

Explain why the car's total energy decreases. (3 marks)

....

....

....

Score   /10

For more help on this topic see KS3 Science Revision Guide pages 58–59.

A Choose just one answer: a, b, c or d.

1 Which is the correct equation for speed? (1 mark)

a) speed = $\dfrac{\text{distance}}{\text{time}}$ ○

b) speed = distance × time ○

c) speed = $\dfrac{\text{time}}{\text{distance}}$ ○

d) speed = time + distance ○

2 Which of these are units of speed? (1 mark)

a) grams per second ○

b) metres per second ○

c) seconds per metre ○

d) metres per gram ○

3 If time and speed are known, how can distance be calculated? (1 mark)

a) distance = speed × time ○

b) distance = $\dfrac{\text{speed}}{\text{time}}$ ○

c) distance = $\dfrac{\text{time}}{\text{speed}}$ ○

d) distance = speed − time ○

4 If speed and distance are known, how can time be calculated? (1 mark)

a) time = $\dfrac{\text{distance}}{\text{speed}}$ ○

b) time = distance − speed ○

c) time = $\dfrac{\text{speed}}{\text{distance}}$ ○

d) time = speed × distance ○

5 Which type of graph shows speed? (1 mark)

a) a mass over time graph ○

b) a pressure over distance graph ○

c) a distance over time graph ○

d) a volume over time graph ○

Score /5

B Answer each question.

1 Work out the speed of an object that travels 20m in 10 seconds. (1 mark)

....

2 Calculate the speed of an object, in metres per second, that travels 1km in 30 seconds. (1 mark)

....

3 Calculate the speed of an object, in metres per second, that travels 20km in 5 minutes. (1 mark)

....

4 Calculate how far an object that is moving at 35m/s travels in 30 seconds. (1 mark)

....

5 Calculate how long it takes an object moving at 2m/s to travel 1400m. (1 mark)

....

Score /5

Answer all parts of the question. Use a separate sheet of paper if necessary.

1 The graph below shows the distance a person walks over time.

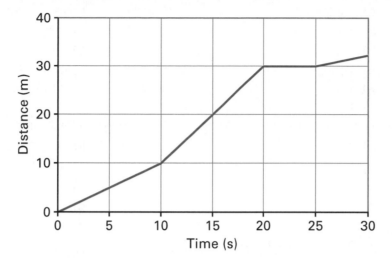

a) When was the person walking fastest? (1 mark)

....

b) Explain your answer to part **a)**. (1 mark)

....

....

c) When was the person standing still? (1 mark)

....

d) Explain your answer to part **c)**. (1 mark)

....

....

e) Describe the shape of the graph above if the following occurred after 30 seconds:
the person started running fast, then slowed down and then stopped. (3 marks)

....

....

....

Score /7

For more help on this topic see KS3 Science Revision Guide pages 62–63.

A Choose just one answer: a, b, c or d.

1 Push and pull are examples of: (1 mark)
- a) electricity ◯
- b) energy ◯
- c) gravity ◯
- d) forces ◯

2 Which are the units of force? (1 mark)
- a) ohms ◯
- b) volts ◯
- c) newtons ◯
- d) joules ◯

3 Which force opposes movement through air? (1 mark)
- a) upthrust ◯
- b) magnetism ◯
- c) conduction ◯
- d) air resistance ◯

4 Which force is produced when travelling along a surface? (1 mark)
- a) gravity ◯
- b) friction ◯
- c) upthrust ◯
- d) Hooke's ◯

5 Which name is given to a turning force around a pivot? (1 mark)
- a) moment ◯
- b) centripetal ◯
- c) centrifugal ◯
- d) gravity ◯

Score /5

B Answer each question.

1 If an object is moving at a constant speed, are the forces on it balanced or unbalanced? (1 mark)

....

2 What will happen to a stationary object which has balanced forces on it? (1 mark)

....

3 Give two examples of non-contact force. (2 marks)

....

4 What is Hooke's law? (2 marks)

....

....

....

5 Describe the balance of forces on an object when it is stretched. (1 mark)

....

....

Score /7

Answer all parts of the question. Use a separate sheet of paper if necessary.

1 The diagram below shows the forces acting on a car.

1200N   1000N

a) Is the car speeding up or slowing down?   (1 mark)

....

b) Explain your answer to part **a)**.   (1 mark)

....

....

c) The driver of the car is keen to reduce the force that opposes the car's movement, so the car can go faster.

What force is the driver trying to reduce?   (1 mark)

....

d) The diagram below shows the same car travelling at the same speed as in the previous diagram. The driver has fitted a spoiler to the car to make it more aerodynamic.

1200N   1100N

Is the spoiler having a positive effect?   (1 mark)

....

e) Explain your answer to part **d)**.   (1 mark)

....

....

f) The car's weight is 7500N. What is the opposing force from the Earth when the car is stationary?   (1 mark)

....

Score   **/6**

For more help on this topic see KS3 Science Revision Guide pages 64–65.

A Choose just one answer: a, b, c or d.

1 Which of the following statements is correct? (1 mark)
   a) pressure is force over the area the force is applied ◯
   b) pressure is mass over the volume the mass is applied ◯
   c) pressure is force over the mass the force is applied ◯
   d) pressure is concentration over the area ◯

2 Which of the following statements is true? (1 mark)
   a) gases and solids are fluids and exert pressure ◯
   b) gases and liquids are fluids and exert pressure ◯
   c) gases and solids are fluids and cannot exert pressure ◯
   d) gases and liquids are fluids and cannot exert pressure ◯

3 Which of these causes atmospheric pressure? (1 mark)
   a) water ◯   b) rock ◯
   c) air ◯   d) mercury ◯

4 Which of the following would decrease the pressure of a fluid? (1 mark)
   a) lowering the temperature ◯
   b) increasing the kinetic energy of the particles ◯
   c) increasing the temperature ◯
   d) increasing the mass of the sample ◯

5 Where is the pressure highest in the ocean? (1 mark)
   a) at the surface ◯
   b) just below the surface ◯
   c) mid-way between the bottom and the surface ◯
   d) at the bottom of the ocean ◯

Score /5

B Answer all parts of each question.

1 a) What happens to atmospheric pressure as you move up through the atmosphere? (1 mark)

....

   b) Explain your answer to part **a)**. (2 marks)

....

2 a) What happens to water pressure as you go deeper beneath the surface? (1 mark)

....

   b) Explain your answer to part **a)**. (2 marks)

....

Score /6

Answer all parts of the questions. Use a separate sheet of paper if necessary.

1 The table below shows the density of different samples of materials.

Sample	Density (kg/m³)
A	240
B	1500
C	700
D	1200

The density of water is 1000kg/m³.

a) Predict which samples will sink and which will float in water by writing the appropriate letters in the table. (2 marks)

Floats	Sinks

b) Explain your answer to part **a)**. (2 marks)

....

....

c) Name the force that causes the samples to float. (1 mark)

....

2 When designing submarines, why is it important to take into account water pressure? (4 marks)

....

....

....

....

....

....

Score   /9

For more help on this topic see KS3 Science Revision Guide pages 66–67.

A — Choose just one answer: a, b, c or d.

1 Which type of wave is a water wave? (1 mark)
- a) longitudinal ◯
- b) electromagnetic ◯
- c) transverse ◯
- d) biochemical ◯

2 Which type of wave is a sound wave? (1 mark)
- a) longitudinal ◯
- b) electromagnetic ◯
- c) transverse ◯
- d) biochemical ◯

3 Which term refers to the number of waves that pass a point in a second? (1 mark)
- a) wavelength ◯
- b) amplitude ◯
- c) waveform ◯
- d) frequency ◯

4 Which term refers to sound waves bouncing off objects? (1 mark)
- a) refraction ◯
- b) amplitude ◯
- c) echo ◯
- d) diffraction ◯

5 Which of these occurs when two waves hit each other? (1 mark)
- a) diffraction ◯
- b) superposition ◯
- c) refraction ◯
- d) echolocation ◯

Score /5

B — Answer each question.

1 Which would have a lower frequency, a low-pitched or a high-pitched sound? (1 mark)

....

2 What is 'auditory range'? (2 marks)

....

....

3 What is 'ultrasound'? (1 mark)

....

....

4 State two uses of ultrasound. (2 marks)

....

....

Score /6

Answer all parts of the questions. Use a separate sheet of paper if necessary.

1 The table below shows an investigation into the speed of sound in different materials.

Medium	Speed of sound (m/s)
Copper	3560
Gold	3240
Glass	5640

a) What conclusion can you draw from this data about the speed of sound in metals and non-metals? (1 mark)

....

b) How reliable is your conclusion? Explain your answer. (2 marks)

....

....

c) Describe how you could improve the reliability of the investigation's results. (2 marks)

....

....

....

2 James is watching a science fiction film.

A spaceship explodes in space and makes a loud sound. James doesn't believe this is scientifically accurate.

Is James correct? Explain your answer. (4 marks)

....

....

....

....

....

Score   /9

For more help on this topic see KS3 Science Revision Guide pages 70–71.

A Choose just one answer: a, b, c or d.

1 Which type of diagram can be used to show the behaviour of light? (1 mark)
a) a ray diagram ○
b) a wave diagram ○
c) a superposition diagram ○
d) a convex diagram ○

2 Which of these travels fastest? (1 mark)
a) sound ○
b) ultrasound ○
c) light ○
d) pressure waves ○

3 What light is reflected by a red object? (1 mark)
a) no light is reflected ○
b) all the light is reflected ○
c) all light except the red light ○
d) only red light is reflected ○

4 What happens to light when it passes from a more dense object to a less dense object? (1 mark)
a) refraction ○
b) superposition ○
c) rarefaction ○
d) nothing ○

5 What type of lens can be used to focus light on a point? (1 mark)
a) a concave lens ○
b) a convex lens ○
c) an opaque lens ○
d) a divergent lens ○

Score /5

B Answer each question.

1 What happens to white light when it is shone through a prism? (2 marks)

2 In specular reflection, what does the angle of incidence equal? (1 mark)

3 What colours of light are reflected by a black object? (1 mark)

4 What colours of light are reflected by a white object? (1 mark)

Score /5

Answer all parts of the questions. Use a separate sheet of paper if necessary.

1 The ray diagrams below show two types of reflection.

A   B

shiny surface   matt or rough surface

a) What type of reflection is shown in each diagram?   (2 marks)

A ....

B ....

b) Which ray diagram would form an image?   (1 mark)

....

c) Why is light not reflected by transparent materials?   (1 mark)

....

2 The diagram below shows a ray of light entering a glass block.

Draw a ray diagram to show how the light would behave as it travelled through the block
and out the other side.   (2 marks)

Glass block

Score   /6

For more help on this topic see KS3 Science Revision Guide pages 72–73.

A — Choose just one answer: a, b, c or d.

1 Electric current is the flow of: (1 mark)
- **a)** ions ○
- **b)** voltage ○
- **c)** charge ○
- **d)** resistance ○

2 Which term describes difference in energy between two different parts of a circuit? (1 mark)
- **a)** resistance difference ○
- **b)** potential difference ○
- **c)** ohmic difference ○
- **d)** amplitudal difference ○

3 Which of these is represented by a straight line in a circuit? (1 mark)
- **a)** a bulb ○
- **b)** an ammeter ○
- **c)** a voltmeter ○
- **d)** a wire ○

4 Which of these is a type of circuit? (1 mark)
- **a)** parallel ○
- **b)** resistive ○
- **c)** ampic ○
- **d)** voltic ○

5 What are the correct units for resistance? (1 mark)
- **a)** volts ○
- **b)** ohms ○
- **c)** amps ○
- **d)** seconds ○

Score /5

B — Answer each question.

1 What is the function of a cell or battery in a circuit? (1 mark)

....

2 What component is used to measure current? (1 mark)

....

3 What component is used to measure voltage? (1 mark)

....

4 What would have the highest resistance, an insulator or a conductor? (1 mark)

....

Score /4

C **Answer all parts of the questions. Use a separate sheet of paper if necessary.**

1 Identify the following components in a circuit. (3 marks)

⊗ ....

⌿ ....

Ⓐ ....

Ⓥ ....

▭ ....

2 The tables show the current in different parts of two circuits, A and B.

Circuit A

Position in circuit	Current (A)
1	4
2	4
3	4

Circuit B

Position in circuit	Current (A)
1	9
2	3
3	6

a) What type of circuit is Circuit A? Explain your answer. (3 marks)

....

....

....

b) What type of circuit is Circuit B? Explain your answer. (3 marks)

....

....

....

....

Score   **/9**

For more help on this topic see KS3 Science Revision Guide pages 76–77.

A Choose just one answer: a, b, c or d.

1 Which of these can be transferred when two insulators are rubbed together? (1 mark)
a) resistance ○
b) waves ○
c) magnetism ○
d) electrons ○

2 What will happen if two negatively charged objects are brought close together? (1 mark)
a) they will attract each other ○
b) they will have no effect on each other ○
c) they will explode ○
d) they will repel each other ○

3 A magnet produces: (1 mark)
a) a magnetic field ○
b) static electricity ○
c) magnetic waves ○
d) a magnetic beam ○

4 Which of the following materials is magnetic? (1 mark)
a) aluminium ○
b) iron ○
c) water ○
d) plastic ○

5 What do compasses point towards? (1 mark)
a) magnetic north ○
b) magnetic south ○
c) magnetic east ○
d) magnetic west ○

Score /5

B Answer each question.

1 When would two magnets repel each other? (2 marks)

....

....

2 When would two magnets attract each other? (2 marks)

....

....

3 Describe how an electromagnet can be produced. (3 marks)

....

....

....

Score /7

1   The diagram below shows a bar magnet.

    a)   Label the poles on the bar magnet.   (1 mark)

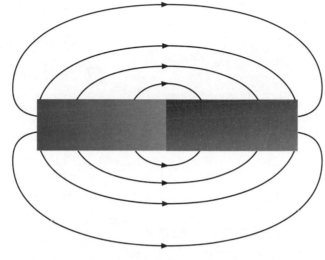

    b)   The bar magnet was brought into contact with different materials.

         Place a tick in the table below if the material would be attracted to the magnet and a cross
if it would not be attracted.   (2 marks)

Material	Attracted
Steel	
Cotton	
Nickel	
Paper	

    c)   Explain why it would be difficult to navigate using a compass if you were carrying a
bar magnet in your pocket.   (3 marks)

....

....

....

2   Describe how a DC motor can be made.   (4 marks)

....

....

....

....

Score   **/10**

For more help on this topic see KS3 Science Revision Guide pages 78–79.

A — Choose just one answer: a, b, c or d.

1 During a physical change of state, what happens to the mass of the matter? (1 mark)
a) it stays the same ○
b) it decreases ○
c) it increases ○
d) it increases then decreases ○

2 What change of state occurs when a solid becomes a liquid? (1 mark)
a) sublimation ○
b) boiling ○
c) freezing ○
d) melting ○

3 What term describes the mass of a substance in a given volume? (1 mark)
a) upthrust ○
b) weight ○
c) absorbance ○
d) density ○

4 Which process occurs when a solute and a solvent form a solution? (1 mark)
a) sublimation ○
b) dissolving ○
c) absorbing ○
d) freezing ○

5 The name given to the process where a solid turns straight into a gas is: (1 mark)
a) boiling ○
b) sublimation ○
c) diffusion ○
d) melting ○

Score /5

B — Answer each question.

1 During what type of change are atoms rearranged to form new compounds and molecules? (1 mark)

....

2 During what type of change do the atoms of matter stay the same but the physical state of the matter changes? (1 mark)

....

3 What term describes the random movement of particles in a fluid? (1 mark)

....

4 Which states of matter are relatively incompressible? (2 marks)

....

5 What state of matter is compressible? (1 mark)

....

Score /6

Answer all parts of the question. Use a separate sheet of paper if necessary.

1 Javed and Tracy carry out an investigation.

They add a liquid solvent to a solid solute. The solute disappears and a liquid is left.

Javed thinks this is an example of melting as the solid disappeared and a liquid was left.

a) Is Javed correct? (1 mark)

....

b) Explain your answer to part **a)**. (2 marks)

....

....

c) Javed and Tracy leave the liquid overnight in a warm room. When they return the liquid has disappeared and the original solid has been left behind.

What process has occurred? (1 mark)

....

Javed and Tracy record the mass of the solute and the mass of the solvent. They also record the mass of the solute at the end of the experiment. Their results are shown below.

	Mass (g)
Solute	5
Solvent	15
Final solute	5

d) What was the mass of the:

i) solution? (1 mark)

....

ii) the liquid lost overnight? (1 mark)

....

e) If this experiment was done in a sealed container, the liquid could be re-formed from the gas. Explain how this could be done. (2 marks)

....

....

Score **/8**

For more help on this topic see KS3 Science Revision Guide pages 82–83.

A Choose just one answer: a, b, c or d.

1 What happens to the kinetic energy of particles when they are heated? **(1 mark)**
a) it increases ○
b) it decreases ○
c) it stays the same ○
d) it increases then decreases ○

2 What happens to the kinetic energy of particles when they are cooled? **(1 mark)**
a) it increases ○
b) it decreases ○
c) it stays the same ○
d) it increases then decreases ○

3 What type of energy is stored in fuels? **(1 mark)**
a) kinetic energy ○
b) light energy ○
c) elastic potential energy ○
d) chemical energy ○

4 What substance is formed when water freezes? **(1 mark)**
a) steam ○
b) hydrogen ○
c) oxygen ○
d) ice ○

5 In which of these states are particles least packed together? **(1 mark)**
a) solid ○
b) liquid ○
c) gas ○
d) ice ○

Score   /5

B Answer each question.

1 How can stored energy be released from fuels? **(1 mark)**

....

2 Explain why ice floats on liquid water. **(3 marks)**

....

....

....

....

3 What process releases energy from food? **(1 mark)**

....

Score   /5

1 Write the numbers 1–3 next to the descriptions of particles, with 1 being the most dense and 3 being the least dense. (2 marks)

Particles very close together	
Particles very far apart	
Particles quite close together	

2 Sara and Nam investigate the properties of water. They cool a small sample of water to −1°C.

a) What substance is produced by cooling the water to this temperature? (1 mark)

....

b) Sara and Nam drop the substance they have produced into a large tank of liquid water.

Describe what happens. (1 mark)

....

c) They leave their experiment for several hours and then return.

Use the words below to fill in the blanks to describe what will have happened when they return.

Each word can be used more than once. (3 marks)

ice   **energy**   **particles**   **faster**   **liquid**   **bonds**

The .... in the .... will gain .... This will cause the

.... to move faster, breaking the .... between the particles. The

.... will melt, forming .... water.

Score **/7**

For more help on this topic see KS3 Science Revision Guide pages 84–85.

A Choose just one answer: a, b, c or d.

1 Which of these attracts objects to the surface of the Earth? (1 mark)
a) magnetism ○
b) repulsion ○
c) gravity ○
d) strong forces ○

2 The Sun is a: (1 mark)
a) galaxy ○   b) planet ○
c) comet ○   d) star ○

3 What is a light year? (1 mark)
a) a unit of distance ○
b) a unit of time ○
c) a unit of mass ○
d) a unit of volume ○

4 Which term describes the Sun and the planets around it? (1 mark)
a) galaxy ○
b) universe ○
c) nebula ○
d) solar system ○

5 The Milky Way is an example of a: (1 mark)
a) solar system ○
b) universe ○
c) galaxy ○
d) black hole ○

Score /5

B Answer each question.

1 A car with a mass of 1500kg is on the Earth's surface. What is its weight? (1 mark)

Gravitational field strength on Earth = 10N/kg

weight = mass × gravitational field strength

2 A lunar rover with a mass of 360kg is on the surface of the Moon. What is its weight? (1 mark)

Gravitational field strength on the Moon = 1.6N/kg

weight = mass × gravitational field strength

3 Explain why it is winter in the northern hemisphere whilst at the same time of year it is summer in the southern hemisphere. (3 marks)

4 What force is caused by gravity? (1 mark)

Score /6

Answer all parts of the question. Use a separate sheet of paper if necessary.

1 The graph below shows the gravitational field strength on different planets.

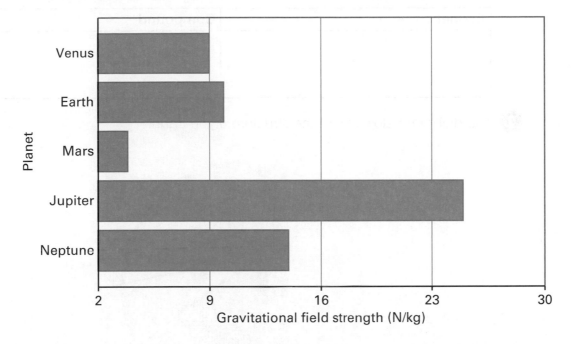

a) Put the planets in order of their gravitational field strength, starting with the largest gravitational field strength. (1 mark)

....

....

b) Which planet has:

i) the smallest mass? (1 mark)

....

ii) the largest mass? (1 mark)

....

c) Explain your answer to part **b)**. (2 marks)

....

....

....

....

Score   /5

For more help on this topic see KS3 Science Revision Guide pages 86–87.

1 Complete the table to show which of these substances are elements and which are compounds. **(3 marks)**

HCl   Ba   CO₂   NaF   O₂   Ru

Element	Compound

2 The diagram below shows the structure of the Earth.

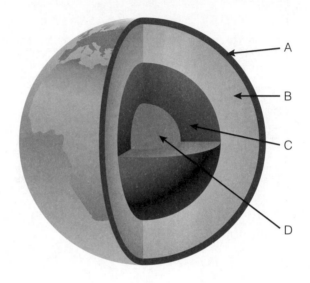

a) Identify layers A–D. **(4 marks)**

A ....

B ....

C ....

D ....

b) Complete the table below to show which of the layers identified in part **a)** are solid and which are liquid. **(2 marks)**

Solid	Liquid

3 Distance can be measured using different units.

Complete the table below using the following options to show
the most suitable unit for each example. **(2 marks)**

metres   mm   km   light years

Distance	Most appropriate unit
The diameter of a galaxy	
The distance between two cities	
The width of a school sports field	
The length of an insect	

4 The diagram below shows a simplified version of the carbon cycle.

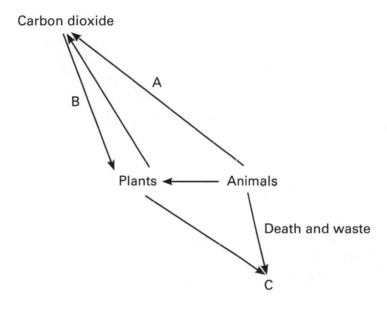

Identify the processes labelled A–C. **(3 marks)**

A ....

B ....

C ....

5 The diagram below shows a plant and an animal cell.

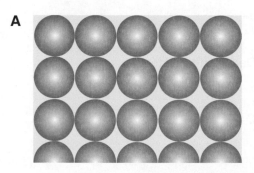

A plant cell from a leaf

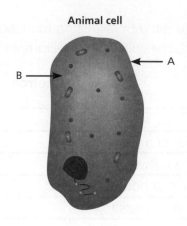

Animal cell

Give the names of the parts labelled A–D. **(4 marks)**

A ....

B ....

C ....

D ....

6 The diagrams below show the particles in three states of matter.

A

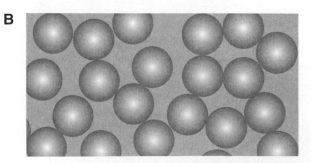

B

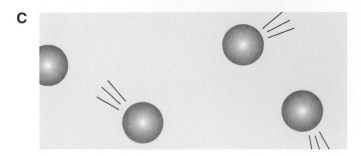

C

a) Name each of the states. **(3 marks)**

A ....

B ....

C ....

b) Complete the table below to show the properties of the different states of matter. **(3 marks)**

State of matter	Shape	Volume	Compressible
			Easily compressed
	Fixed shape		
Liquid		Fixed volume	

7 The diagram below shows a complete circuit.

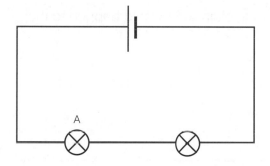

a) What type of circuit is this? **(1 mark)**

....

b) What is component A? **(1 mark)**

....

c) The bulb is not lit in the circuit below. Explain why. **(2 marks)**

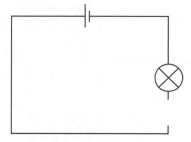

....

....

8 The diagram below shows a simple food chain.

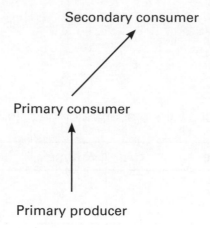

Secondary consumer

Primary consumer

Primary producer

a) What process does a primary producer use to produce energy? **(1 mark)**

....

b) What would be the effect on the primary producer if it was unable to get enough water? **(1 mark)**

....

c) Use your answer to **b)** to explain what would happen to the:

i) primary consumer **(1 mark)**

....

....

ii) secondary consumer **(1 mark)**

....

....

9 Write the missing information in the table using the options given. Each metal should only be used once. **(3 marks)**

Silver   Copper   Zinc   Potassium

Metal	Description
	Can be extracted from its ore using carbon and is more reactive than iron
	Doesn't form an ore
	Less reactive than lead but more reactive than silver
	Most reactive metal in this table

10 a) Complete the table by writing the words 'high' or 'low' to show the concentrations of oxygen and carbon dioxide in inhaled and exhaled air.   **(2 marks)**

	Oxygen	Carbon dioxide
Inhaled		
Exhaled		

b) Explain your answer to part **a)**.   **(2 marks)**

....

....

....

11 A cell was placed in a solution. The graphs show the concentrations of five different substances (A, B, C, D and E) in the cell and in the solution. All of the substances were able to move through the cell membrane by diffusion.

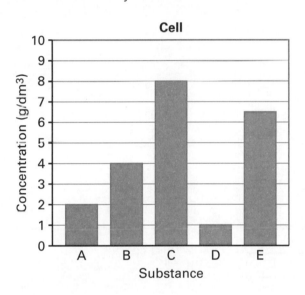

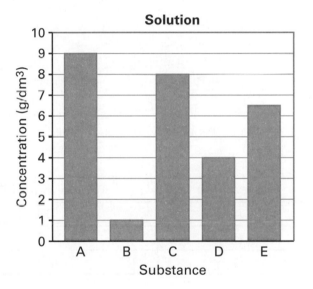

Complete the table with the letters A, B, C, D and E to show the direction of movement of the five different substances.   **(3 marks)**

Substance moves from the solution into the cell	Substance moves from the cell into the solution	There is no net movement of substance

12 Eggs are being fried in a pan on a stove.

a) What type of energy is being transferred to the eggs? **(1 mark)**

....

The stove is switched off.

b) Do the eggs have a higher or lower energy than before they were cooked?

Explain your answer. **(2 marks)**

....

....

c) Describe what would happen if the eggs were left for a long period of time without being reheated. **(3 marks)**

....

....

....

13 Peter and Anuja carry out an investigation into pressure in gases. The table below shows the temperature of the three samples of gas they study. The gas was contained in a fixed volume container.

Gas sample	Temperature (°C)
A	100
B	200
C	120

a) Which sample would have the highest pressure?

Explain your answer. **(3 marks)**

....

....

....

Peter and Anuja decide to conduct a different experiment. This time the gas is kept at the same temperature but the volume of the container is changed.

Gas sample	Volume of container (cm³)
A	50
B	90
C	20

b) Which sample would have the highest pressure?

Explain your answer. **(3 marks)**

....

....

....

14 Put the following statements in order to describe how a singer's voice is recorded. Write the numbers 1–6 in each box. The first one has been done for you. **(3 marks)**

The vibrations of the diaphragm are converted to an electrical signal ☐

The singer's vocal chords vibrate 1

The sound waves hit the diaphragm of the microphone ☐

A longitudinal sound wave is produced ☐

The diaphragm of the microphone vibrates ☐

The electrical signal is transmitted to the recording device down a wire ☐

15 Match the words below to the pictures of the levels of organisation of living organisms. **(2 marks)**

tissue   **organism**   **organ system**   **organ**

A   B   C   D

A ....

B ....

C ....

D ....

16 An experiment is carried out into substance A. A molecule of substance A consists of carbon, hydrogen and oxygen.

a) Use the following words to complete the sentence below. You can use each word more than once. **(1 mark)**

elements   compound

Carbon, hydrogen and oxygen are all .... As substance A is made up of

three different .... joined together it is a ....

Two experiments were carried out on substance A.

Experiment A: substance A was heated until it changed from a liquid to a gas.

Experiment B: substance A was reacted with another substance.

b) Use the following terms to describe experiment A and experiment B. **(1 mark)**

chemical reaction   change of state

Experiment A

....

Experiment B

....

c) The pie chart below shows the percentages of carbon, hydrogen and oxygen in substance A. Complete the other pie chart to show the percentages of carbon, hydrogen and oxygen after experiment A. **(1 mark)**

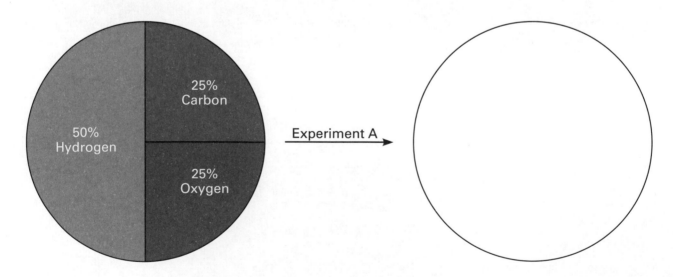

17 The following statements describe the passage of air through the respiratory system of a human. Put them in the correct order by writing the numbers 1–5 in each box. The first one has been done for you. **(2 marks)**

Air travels through the bronchi ☐

Air enters the mouth 1

Air enters the alveoli ☐

Air travels down the trachea ☐

Air enters the bronchioles in the lungs ☐

18 The diagram shows a piece of equipment that is used to separate mixtures.

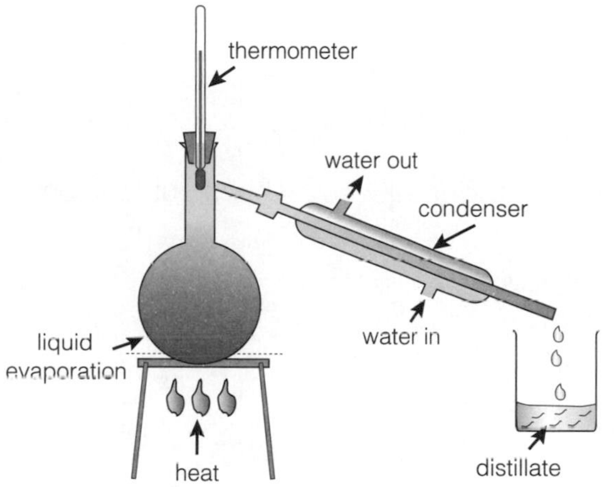

a) Name the separation process that this apparatus is used for. **(1 mark)**

....

b) Describe the function of:

i) the heat **(2 marks)**

....

....

ii) the condenser **(2 marks)**

....

....

19 The photograph below shows an object that can be used in a physics experiment to investigate the colours of light.

a) What is this object called? **(1 mark)**

....

b) Give the order of the colours produced when white light is shone through this object. **(2 marks)**

....

....

....

c) Explain why white light splits into these colours when it is shone through the object. **(2 marks)**

....

....

....

In order to observe these colours, light must enter our eyes and hit the retina.

d) What part of the eye focuses the light on the retina? **(1 mark)**

....

e) Describe what happens when light hits the retina. **(2 marks)**

....

....

....

20 The statements below refer to breathing in and breathing out. Write 'I' in the boxes next to the statements that correspond to breathing in and 'O' in the boxes next to the statements referring to breathing out. **(4 marks)**

The external intercostal muscles contract and pull the ribs up and out ☐

The pressure in the lungs increases ☐

The pressure in the lungs decreases ☐

The lungs increase in volume ☐

The diaphragm relaxes and raises ☐

The diaphragm contracts and flattens ☐

The external intercostal muscles relax, the ribs move down and in ☐

The lungs decrease in volume ☐

21 Liam investigates separating mixtures using evaporation.

a) What type of mixtures can be separated by evaporation? **(1 mark)**

....

Liam sets his experiment and leaves his mixtures in open containers for an hour.

b) Why is it important to leave the mixtures in open containers? **(2 marks)**

....

....

c) After an hour, Liam returns to his experiment. He is disappointed to see that very little evaporation has occurred.

How could he speed up the rate of evaporation in his experiment? **(1 mark)**

....

22 Nabeela carries out an investigation into making saltwater. She dissolves salt into water.

a) Complete the table below to show how Nabeela's experiment was set up. **(3 marks)**

	Solute / Solvent / Solution	Mass (g)
Salt		5
Water		100
Saltwater		

b) Nabeela wants to separate the salt from the saltwater. Could she use filtration to do this?

Explain your answer. **(3 marks)**

....

....

....

23 The diagrams below show the male and female reproductive systems of humans.

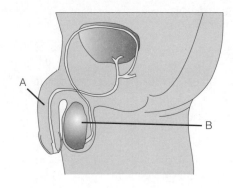 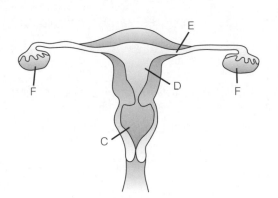

Identify the labelled parts of the reproductive systems. **(3 marks)**

A ....

B ....

C ....

D ....

E ....

F ....

24 During exercise, the cells in the muscle do not receive enough oxygen to respire aerobically.

a) How are the muscle cells still able to release energy? **(1 mark)**

....

b) When exercising, a burning sensation can occur in the muscles.

What product of anaerobic respiration causes this burning sensation? **(1 mark)**

....

c) Apart from the above, what is another disadvantage to muscle cells respiring without oxygen? **(1 mark)**

....

d) Explain how exercise:

    i) benefits the gas exchange system. **(1 mark)**

....

....

    ii) reduces the risk of obesity. **(1 mark)**

....

....

e) Explain why someone with asthma finds it difficult to exercise. **(2 marks)**

....

....

....

....

25 Pawel and Jenny are investigating different types of igneous rocks. Their notes are shown below.

> **Rock A: made up of small crystals**
> **Rock B: made up of large crystals**

a) Explain how igneous rocks are formed. **(2 marks)**

....

....

....

b) Which rock, A or B, is an intrusive igneous rock?

Explain your answer. **(3 marks)**

....

....

....

c) Which rock, A or B, is an extrusive igneous rock?

Explain your answer. **(3 marks)**

....

....

....

26 a) What type of magnet is shown in the diagram below? **(1 mark)**

....

b) What is the shape of the magnetic field produced by this magnet? **(1 mark)**

....

c) When would the magnetic field be produced? **(1 mark)**

....

d) Give two ways in which the strength of the magnetic field could be increased. **(2 marks)**

....

....

27 DDT is a pesticide that has been shown to affect food chains. Number the statements below in the correct order to show the effect of DDT on a food chain. Write the numbers 1–8 in the boxes. The first one has been done for you. **(3 marks)**

DDT is sprayed on crops
| 1 |

Small birds feed on many insects
| |

The population of hawks decreases
| |

Insects feed on the crops and take in DDT
| |

Due to DDT, the hawks' eggs are very fragile and break
| |

Hawks feed on many small birds
| |

DDT accumulates in the small birds' bodies
| |

DDT accumulates in the hawks' bodies
| |

28 A population of birds feeds on nuts. Some birds have small beaks whilst other birds have large beaks. The birds with larger beaks are able to break open the nuts in order to eat them.

a) What term can be used to describe the fact that different members of the species have different beak sizes? **(1 mark)**

....

b) What are the birds competing for? **(1 mark)**

....

c) Why will the birds with the larger beaks be more likely to survive? **(2 marks)**

....

....

d) The offspring of the birds with large beaks are more likely to survive. Over a long period of time all the birds in the population would have large beaks.

What is the name given to the process described? **(1 mark)**

....

29 When fossil fuels are burned, non-metal oxides are produced. The non-metals dissolve in water in the atmosphere and fall as rain.

a) Name the type of substance formed when a non-metal oxide dissolves in water. **(1 mark)**

....

b) Explain why this rain might cause damage to buildings. **(2 marks)**

....

....

....

c) What colour would this rain turn universal indicator solution? **(1 mark)**

....

d) What chemical reaction could occur if the rain came into contact with a metal? **(1 mark)**

....

30 A boat travels 100 metres in 6 seconds.

a) Calculate the speed of the boat. **(1 mark)**

....

b) Calculate how far the boat would travel if it maintained this speed for 20 seconds. **(1 mark)**

....

c) Calculate how long it would take the boat to travel 2km if it maintained its speed. **(1 mark)**

....

d) What force opposes the boat's movement through the water? **(1 mark)**

....

e) The table below shows the distance travelled by another boat over a period of time.

Time (s)	Distance (m)
0	0
20	250
40	320
60	700

On the grid below draw a distance–time graph to represent this data. **(4 marks)**

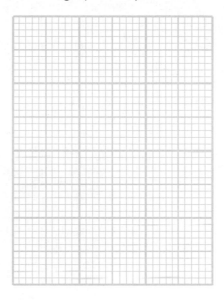

31 Dinosaurs lived millions of years ago.

a) Explain why dinosaurs are said to be extinct. **(1 mark)**

....

b) Scientists are trying to ensure that no species currently alive become extinct.

What term can be used to describe what they are attempting? **(1 mark)**

....

c) Gene banks are one method of preventing further extinctions.

Explain why dinosaurs cannot be brought back from extinction using gene banks. **(2 marks)**

....

....

....

....

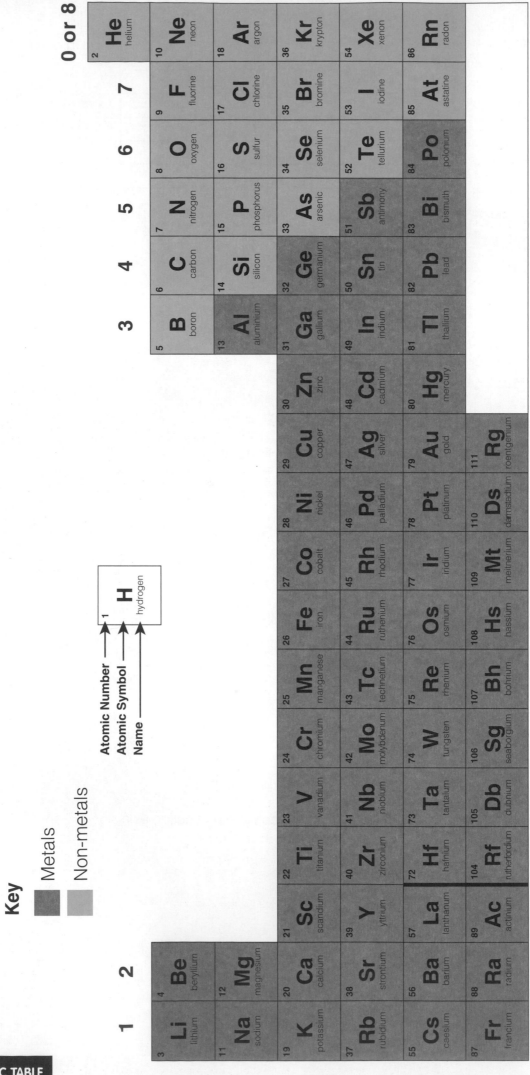